VAN NOSTRAND REINHOLD RETURN POLICY

Thank you for purchasing this VNR software package.
  Before breaking the software seal, please read through the step-by-step documentation to determine if the software is right for your particular needs. If, after reading the documentation, you decide this software is not appropriate for your purposes, you may return this package to VNR for a full refund within 15 days. Unless the disks are defective, this software package will not be accepted for refund if the software seal is broken.

COMPUTER METHODS IN
THE GEOSCIENCES

Daniel F. Merriam, Series Editor

Computer Applications in Petroleum Geology
    Joseph E. Robinson
Graphic Display of Two- and Three-Dimensional Markov
Computer Models in Geology
    Cunshan Lin and John W. Harbaugh
Image Processing of Geological Data
    Andrea G. Fabbri
Contouring Geologic Surfaces with the Computer
    Thomas A. Jones, David E. Hamilton,
    and Carlton R. Johnson
Exploration-Geochemical Data Analysis with the IBM PC
    George S. Koch, Jr. (with programs on diskettes)

Related Titles

Statistical Analysis in Geology
    John M. Cubitt and Stephen Henley (eds.)
Cluster Analysis for Researchers
    H. Charles Romberg
Analysis of Messy Data, Volume 1: Designed Experiments
    George A. Milliken and Dallas E. Johnson

EXPLORATION-GEOCHEMICAL DATA ANALYSIS WITH THE IBM PC

(with programs on diskettes)

GEORGE S. KOCH, JR., University of Georgia

 VAN NOSTRAND REINHOLD COMPANY

New York

Manufactured in the United States of America.

Published by Van Nostrand Reinhold Company Inc.
115 Fifth Avenue
New York, New York 10003

Van Nostrand Reinhold Company Limited
Molly Millars Lane
Wokingham, Berkshire RG11 2PY, England

Van Nostrand Reinhold
480 La Trobe Street
Melbourne, Victoria 3000, Australia

Macmillan of Canada
Division of Canada Publishing Corporation
164 Commander Boulevard
Agincourt, Ontario MIS 3C7, Canada

15 14 13 12 11 10 9 8 7 6 5 4 3 2 1

Library of Congress Cataloging-in-Publication Data
Koch, George S., Jr.
  Exploration-geochemical data analysis with the IBM/PC.
  (Computer methods in the geosciences)
  Includes index.
  1. Geochemical prospecting — Computer programs.   2. IBM Personal
Computer — Programming.   3. BASIC (Computer program language)
I. Title. II. Series.
TN270.K58   1987   622'.13'0285   86-7807
ISBN-13: 978-1-4612-9171-8   e-ISBN-13: 978-1-4613-1973-3
DOI:10.1007/978-1-4613-1973-3

Contents

Disclaimer

Neither Van Nostrand Reinhold Company nor the author nor any employer of the author shall be liable for any special, indirect, consequential, incidental or other similar damages suffered by the user or any third party, including, without limitation, damages for loss of profits or business or damages resulting from use or performance of the software, the documentation, or any information supplied by the software or documentation, whether in contract or in tort, even if Van Nostrand Reinhold or its authorized representative has been advised of the possibility of such damages; and Van Nostrand Reinhold shall not be liable for any expenses, claims or suits arising out of or relating to any of the foregoing.

User Assistance and Information

Any problems, comments, or suggestions regarding these problems should be directed to Dr. George S. Koch, Jr., Department of Geology, University of Georgia, Athens, GA 30602.

Series Editor's Foreword

Here is another contribution in the continuing series on Computer Methods in the Geosciences. As its title suggests, this volume will be of interest to explorational geochemists who want to analyze their own data on a personal computer (PC). To make it easy for the user, the programs and two trial data sets are provided on the accompanying diskettes. And, by supplying the diskettes, another first is accomplished for the series: instant involvement and interaction for the user.

Although other books in the series have provided listings of computer programs, *Exploration-Geochemical Data Analysis with the IBM PC* is the first to supply diskettes. The diskettes, along with the instructions outlined in the text, eliminate the bother (and errors) of putting the programs in manually. The suite of programs — for handling and sorting data files; computing and displaying summary statistics; and working with logarithms, geochemical thresholds, and regression — will give geochemists a good repertoire for geochemical exploration data analysis. The diskettes are easy to use and have been tested thoroughly.

The PC has changed the work habits of those in the earth sciences. Now, with a powerful analytical tool readily available, easy to use, and economical, the worker can manipulate and analyze the data in numerous ways impossible just a few years ago or possible only with considerable difficulty and effort. The PC has revolutionized those aspects of earth science such as geochemistry where large databases are necessary and available. This guide will help the exploration geochemist (and others who analyze geochemical data) through entering, organizing, editing, and analyzing his or her data.

George Koch has many years of experience and was one of the first to apply computers in his discipline. His books with Dick Link on *Statistical Analysis of Geological Data* (vol. 1, 1970 and vol. 2, 1971, John Wiley & Sons) and with Dick Link and Jack Schuenemeyer on *Computer Programs for Geology* (1972, Artronic Information Systems, Inc.) stand as testimonials to his expertise and background in the subject.

Dr. Koch states in his preface that *Exploration-Geochemical Data Analysis with the IBM PC* was written specifically for exploration geochemists and contains a minimum of jargon and extraneous materials; thus, this contribution will be a welcome addition for all who want to do their own thing. This book is the fifth in the open-ended series initiated in 1982 to (1) promote geomathematics in plain English, (2) introduce the reader to the subject, and (3) keep the geologic public informed of the latest developments. It is apparent that George Koch's contribution admirably fulfills all of the objectives.

DANIEL F. MERRIAM

Preface

This book presents an integrated system of computer programs to use on an IBM personal computer for the analysis of exploration geochemical data. Written in the standard Microsoft BASIC language, the programs can be run on other computers (with program modifications if necessary) or rewritten for other languages. The programs, together with two example data files, are listed on the floppy disks accompanying this book. They are ready for you to start to work, guided by the tutorials.

I have written this book specifically for exploration geochemists and others who analyze exploration geochemical data. Because exploration geochemistry is a field discipline generally practiced away from mainframe computers, microcomputers are particularly well-suited for treating the resulting data. Additional benefits of microcomputers in any case are the following: economy, personal control, and reliability. The programs in this book are written so that data can be passed readily between mainframe computers and microcomputers in order to fully exploit both computing modes if they are available.

Exploration geochemists do not need to understand the intricacies of computers, but they need to extract meaning from their data. It seldom works as well to simply hand over your data to an outsider; personal intervention is needed, at least in some part of the process. For example, you may want to try out a procedure on small data sets to increase your understanding of it or to see how it works on your data. (Similarly, many of us make rough income-tax calculations even though we employ a professional for the final return.)

I intend the computer programs in this book to be easy to learn, foolproof to use, and convenient to modify if you wish. They have clear running instructions written in standard English (rather than in computer jargon). They are designed to check the validity of every keyboard entry and to tell you how to correct typing mistakes, for instance in keying in an alphabetic character when a numeric one is needed. To make the programs easy to modify, they are written in standard Microsoft BASIC (the most widely used language for microcomputers) structured with subroutines, provided with abundant "remark" statements, and accompanied by lists of variable names and definitions. Error messages are truly informative and are written in standard English.

For some purposes, I find large scale, general purpose programs useful as adjuncts to the programs in this book. Therefore, I have included a way to link with the system in this book files constructed with Lotus 1-2-3 and with WordStar.

Some of these programs are adapted from an earlier book, *Computer Programs for Geology,* by myself, R. F. Link, and J. H. Schuenemeyer (1972, Artronic Information Systems, New York). Program 8-1 is by A. T. Miesch; the source is cited in Chapter 8.

Demetrios Papacharalampos wrote programs 2-2, 2-3, 7-1, and 10-1, and rewrote all the others from my early drafts. Without his careful effort, this book would not have been completed.

I remain responsible for any errors that remain in the programs.

GEORGE S. KOCH, JR.

CHAPTER **1**

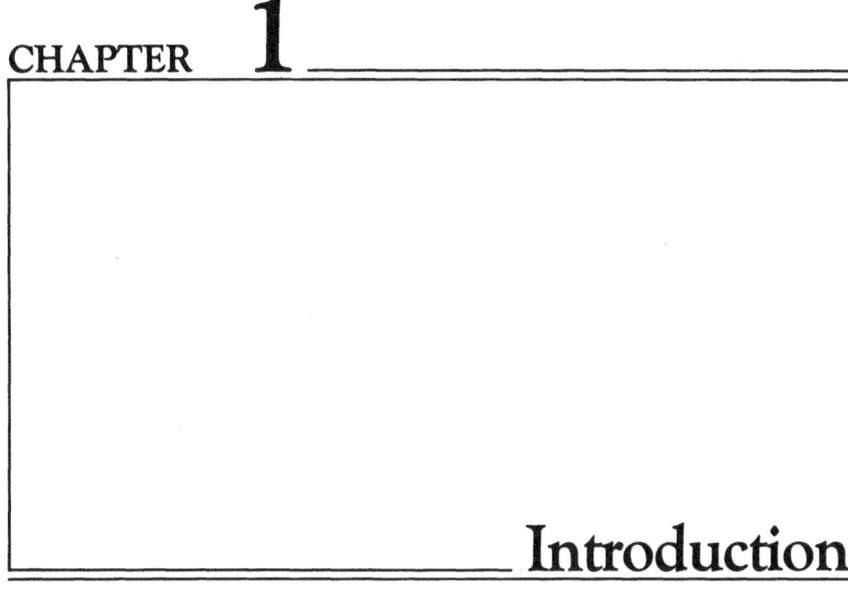

Introduction

This chapter tells you what exploration-geochemical data analysis means to me, why you may want to do your own analysis, what I assume you know before you start using this book and its accompanying programs, and how to get started. Before going any further, following are assumptions I make:

1. You have access to an IBM PC, with at least 256K of dynamic memory and one disk drive, or a larger IBM microcomputer (IBM AT or IBM XT), or a compatible computer made by another vendor. Your microcomputer uses an Epson Spectrum LX-80 or compatible printer.

2. The computer is equipped with the DOS 2.0 (or newer) operating system.

3. You know how to turn on the computer, find your way around the keyboard, and use the DOS operating system to format diskettes, save files, list files, and copy files from one diskette to another.

4. You have at least a few floppy diskettes and know how to take care of them to protect them from physical damage.

EXPLORATION-GEOCHEMICAL DATA ANALYSIS DEFINED

Data are "facts or information to be used as a basis for discussing or deciding something;"[1] *to analyze* is "to separate (a substance etc.) into its parts in order to

1

identify it or study its structure."[2] These dictionary definitions pertain to our subject. We might summarize them by saying that exploration-geochemical data analysis consists of drawing conclusions from the data collected by exploration geochemists. Using a computer, there are several steps or tasks to do:

1. Design an exploration geochemical survey.

2. Collect data according to the survey design.

3. Enter data into a computer.

4. Organize the data in an orderly way.

5. Edit the data to remove erroneous information.

6. Perform statistical analyses.

7. Draw conclusions.

Although not all these tasks may be done in a particular survey and not in exactly the order listed here, the list gives us a general structure. In this book, I concentrate on steps 3 through 6.

Entering data into a computer may be totally organized and mechanized, with field data collected on computer-readable media such as punched cards and analyses provided in computer-readable form from instruments. At the other extreme, the data may be available only in written or printed records that need to be keyed into a computer. In this book, I assume the latter situation.

Organizing data in orderly ways facilitates analysis, and the pattern of organization itself suggests modes of analysis and geochemical patterns in the data.

STRATEGIES

Some effective strategies in the analysis of data follow.

1. Spend as much time as necessary to construct consistent and carefully edited data files. The time spent generally will be a larger part of any particular data analysis than you would like and also more than you spend on any other part. Some characteristics of well-constructed data files are these:

They contain enough information (header lines) preceding the data lines (data records) so that their identity is readily apparent from the file itself.

The variables are arranged systematically in a convenient form that is easy to modify as needed.

The files are constructed through processes that minimize keypunching errors or other errors in entering the data.

The files are systematically edited and the editing procedures are recorded; comments explaining the purposes of the editing are included.

The files are stored as standard ASCII or BASIC files (or in some other standard form) so that their content will not be lost if the system with which they were created fails or is not maintained. More than one copy is saved. At least one of these copies is saved as a backup that can be changed only through a carefully devised formal procedure.

2. Think about the geologic meaning of your data and pose some geologically interesting questions to be answered by a data analysis. Choose some simple questions that can be answered by straightforward procedures, the significance of which you understand.

3. Process your files with a linked system of computer programs that you can modify. Have each program do one or a few simple steps that you clearly understand to provide output, all of which you understand.

4. After perhaps some exploratory data analysis, systematically process the data files with a planned series of steps.

SAMPLE DATA FILES

Two sample data files are provided so that you can practice using the programs to get the results listed in this book. The files are for two sets of exploration-geochemical data.

The first file (Table 1.1), NORWAY, consists of data from 25 stream-sediment samples collected by the Norwegian Geological Survey; Howarth and Sinding-Larsen use these data to illustrate a thorough discussion of multivariate statistical analysis.[3] Figure 1.1 locates the sample points.

The second file (Table 1.2), BUTTE, consists of data from 265 stations in the state of Montana in the northwestern United States. The data are part of a set that the Los Alamos National Laboratory collected in the Butte two-degree quadrangle of the National Topographic Map Series in the National Uranium Resource Evaluation program of the U.S. Department of Energy. Broxton describes the entire data file.[4]

Table 1.2 lists the five lines of information describing the file and the data from the first 13 stations; these 18 lines form the file XBUTTE on the diskette. There are six chemical elements, selected from 53 in the original data file.

Table 1.3 lists the names of data files used in this book to illustrate program operation and also the names of the other files on the diskette.

WHAT ALL THE PROGRAMS IN THIS BOOK HAVE IN COMMON

All these programs were developed together; each consists of particular instructions for the program tasks and utility subroutines that are used by more than one

TABLE 1.1
Listing of Data File NORWAY Containing Sample Data for
25 Stream-sediment Samples from Norway

Sample No.	Zn ppm	Fe pct	Mn ppm	Cd ppm	Cu ppm	Pb ppm
1	24	1.08	330	0.4	7	5
2	25	1.18	420	0.3	9	7
3	42	2.06	910	0.6	12	6
4	50	1.73	700	0.5	15	9
5	52	1.74	690	0.5	18	10
6	29	1.06	510	0.6	10	8
7	26	1.08	530	0.5	10	7
8	23	0.93	260	0.4	7	7
9	89	1.84	670	0.9	32	6
10	72	3.35	530	0.5	11	8
11	31	1.51	350	0.4	11	4
12	115	1.59	650	1.2	37	6
13	535	2.59	960	3.5	350	9
14	48	1.71	570	0.3	17	9
15	1010	2.71	1070	5.6	590	9
16	560	3.05	450	3.1	490	8
17	48	1.43	590	0.4	8	8
18	44	1.70	710	0.4	11	16
19	36	1.28	410	0.2	7	10
20	33	1.94	490	1.0	20	12
21	45	1.79	260	0.8	27	12
22	118	6.70	1930	1.2	33	17
23	274	6.10	5920	3.1	63	27
24	81	2.62	970	0.9	22	13
25	80	3.61	900	1.0	23	10

Source: From R. J. Howarth and R. Sinding-Larsen, 1983, Multivariate Analysis, in *Handbook of Exploration Geochemistry,* Vol. 2, *Statistics and Data Analysis in Geochemical Prospecting,* ed. R. J. Howarth, Elsevier, Amsterdam, p. 210.

program and are called into the program being run. Diskette 2 contains source code for the programs. You can list a program on the printer by entering BASICA from DOS, loading the program, and entering **LLIST**. Appendix A is a list of error messages.

You run the programs by responding to questions and other prompts that appear on the computer screen. Each response is a single entry, which may be a line of alphabetic and/or numeric information, a response of **Y** for yes or **N** for no, a specified number of alphabetical and/or numeric characters, or a single number. After keying each single entry, press the <ENTER> key.

Because most mistakes in data analysis occur in entering data, I have tried to anticipate these mistakes and to help you correct them.

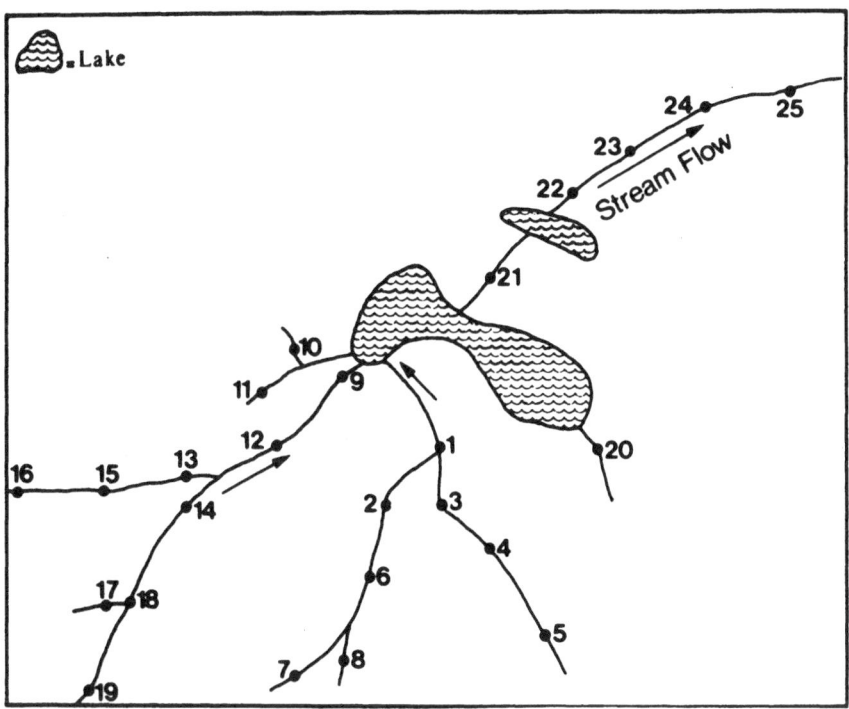

Figure 1.1 Locations of sample points in file NORWAY. (After R. J. Howarth and R. Sinding-Larsen, 1983, Multivariate Analysis, in *Handbook of Exploration Geochemistry*, Vol. 2, *Statistics and Data Analysis in Geochemical Prospecting*, ed. R. J. Howarth, Elsevier, Amsterdam, p. 210.)

Each of your responses must be a definite answer; I do not use *default* responses (entering only a carriage return) because I believe that for beginners this procedure too easily can lead to thoughtless responses or to mistakes caused by accidentally pressing the carriage return key. I have provided messages in clear English to give you a chance to try again in case you make an entry that will not work: for example, entering the word *one* instead of the numeral 1, or entering an *l* (lowercase ell) instead of the numeral 1. Because of the way Microsoft BASICA works, the programs cannot, however, cope with some patterns of repeated errors, as explained further in the last section of this chapter.

Of course, I could not provide for every eventuality. If, for example, you want to be reminded that values of a certain variable cannot exceed some known upper limit, you can modify the programs. That is one reason why the programs are written in standard BASIC language and why the source code is supplied.

TABLE 1-2
Listing of Data File XBUTTE

NURE HSSR DATA FROM BUTTE QUADRANGLE BETWEEN 112 AND 113 WEST AND 46 AND 46.25 NORTH. 265 LOCATIONS. THE FORMAT IS A7,2F9.4,I2,A7,7I7. VARIABLES ARE SAMPLE NUMBER, LATITUDE, LONGITUDE, CONTAMINATION CODE,[a] ROCK CODE,[b] AG, AU, CU, MN, PB, AND ZN. FROM REPORT BY BROXTON, D.E., 1980.

Sample	Latitude	Longitude	Cont.	Rock	AG	AU	CU	MN	PB	ZN
M24164	46.0539	112.6022	3	T-CV---	2	2	32	659	9	95
M25319	46.0144	112.9803	0	T-CV---	2	2	94	554	52	96
M24449	46.2172	112.5694	9	T-CV---	2	2	31	930	41	85
M24196	46.0931	112.5139	3	T-CV---	14	2	68	1517	12	101
M24459	46.2236	112.5250	3	T-CV---	2	2	33	1746	12	34
M24114	46.1833	112.5597	3	T-CV---	2	2	76	427	31	89
M25341	46.0158	112.8114	0	T-CV---	2	2	63	1235	20	71
M24178	46.1317	112.5844	0	T-CV---	2	2	90	584	40	93
M24090	46.1569	112.5403	8	T-CV---	2	2	62	634	28	128
M25313	46.1589	112.9339	0	T-CV---	2	2	821	496	83	153
M24211	46.0408	112.6708	3	T-CV---	2	2	62	495	36	143
M24117	46.2194	112.5625	9	T-CV---	2	2	79	737	44	138
M24170	46.0806	112.5975	3	T-CV---	2	10	39	333	23	69

Notes: This file consists of the first 18 lines of data file BUTTE, which contains example data for 265 stream-sediment samples (extracted from a report by Broxton, 1980). In the file name BUTTES, the letter "S" stands for "short." The fomat is in the notation of the FORTRAN language, used to read the original file.
[a] 1 = none, 2 = mining, 3 = agriculture, 4 = industry, 5 = sewage, 6 = power generation, 7 = urban, 8 = recreation, 9 = other.
[b] The rock codes convey stratigraphic and lithologic information.

TABLE 1-3
List of Files on the Diskettes

Original File Name	Restructured File Name
Part 1. Data files	
EXAMPL-2	
EXAMPL-3	
BUTTE[a]	BUTTE-2
XBUTTE	XBUTTE-2
	XBUTTE-3
YBUTTE	
NORWAY	NORWAY-2
	NORWAY-3
ZINC	ZINC-2
WELKOM	
FRESNI	
CARRIZO	
Part 2. Program files	
MENU.BAS	
SUBROUT.SUB	
P2-1.BAS	
P2-2.BAS	
P2-3.BAS	
P3-1.BAS	
P4-1.BAS	
P5-1.BAS	
P6-1.BAS	
P7-1.BAS	
P8-1.BAS	
P9-1.BAS	
P10-1.BAS	

[a]Because of limited space, file BUTTE is omitted from diskette 1.

HOW TO START

The two diskettes with this book contain duplicate programs. Diskette 1 has the programs in a *compiled* form; diskette 2 has the original source code for the programs so that you can run them using the BASICA *interpreter*. In their compiled form, the programs run much faster, so this is the version to use if you want to run the programs exactly as written. The source code is provided so that you can modify the programs if you wish, presumably recompiling them (using a BASICA compiler, available from a vendor) after they are changed.

You will need to copy the files on the diskettes (called the original diskettes) to working diskettes. There are two reasons for doing this. First, the DOS operating system and BASICA are not provided on the original diskettes, and, second, it is desirable to save the original diskettes so that they will not wear out through repeated use. You can copy either or both. The original diskettes are not write protected; if you are uncertain about the copying procedure you may want to consult your DOS manual and practice on other diskettes.

If you have two disk drives, these instructions will work exactly as written; if you have only one drive, you will need to consult your DOS manual.

1. For each diskette that you want to copy, format a new diskette. To do this, turn on your IBM PC with the DOS operating system in Drive A, enter **FORMAT A:**, and follow the prompts.

2. Copy the files from one of the original diskettes to a formatted diskette. To do this, enter **DISKCOPY A: B:** and follow the prompts. Return the original diskette to the envelope for safekeeping.

3. Now you need to copy files from your DOS diskette to your working diskette. Replace your DOS diskette in drive A and enter the command **B:INSTALL**. Respond to the messages that appear on the screen. On the IBM-PC, the files copied are **SYS, COMMAND.COM**, and **BASICA.COM**. (Diskette 1 has only enough empty space to install DOS version 2.1 or earlier.)

4. The diskette that you made is now ready to use. Repeat Step 1 so that you have a diskette on which to save your data files.

You are now ready to work. For either the diskette with the compiled programs or the one with the source code, place the diskette in drive A. If the computer is still on, perform a warm boot by pressing the <CTRL> <ALT> <DEL> keys at the same time. If the computer is turned off, turn it on and make sure that the drive door is closed. You will see the upper part of Figure 1.2 (surrounded by a border) displayed on the screen. In computer jargon this display is called a *menu*, and it asks you to select among the programs listed. In the figure, I entered the number 1 to select program 2-1, and the program continued with the remaining lines. Operation of this program is described in the next chapter.

You can now continue on with the tutorials in any chapter, using the two sample data files provided, which have been restructured into the file structure used in this system, or you can construct your own data file using Program 2-1.

The system of programs together with the DOS operating system and BASICA take up about 315,000 of the some 360,000 characters available on each floppy diskette. Therefore, you can store one or a few small data files on your program diskette. However, I recommend that you always store data files on another diskette.

TECHNICAL NOTES

These notes record the experience of reviewers and me in running these programs on other than IBM-PC computers and printers and also suggest ways in which you can customize the programs if you wish.

The MS-BASIC interpreters work for the COMPAQ as well as the IBM-PC

```
------------------------------------------------
|                                               |
|   MENU OF PROGRAMS FOR EXPLORATION GEOCHEMISTRY |
|                                               |
|   (1)  To construct a data file (2-1)          |
|   (2)  To correct stations in a data file (2-2)|
|   (3)  To add or delete stations in a data file (2-3)|
|   (4)  To list a data file (3-1)               |
|   (5)  Summary statistics (4-1)                |
|   (6)  To make a histogram (5-1)               |
|   (7)  Working with logarithms (6-1)           |
|   (8)  To sort a data file (7-1)               |
|   (9)  Estimating geochemical thresholds (8-1) |
|   (10) To compare paired observations (9-1)    |
|   (11) To make plots (10-1)                    |
|   (12) Quit and return to DOS                  |
|                                               |
------------------------------------------------

Enter a number between 1 to 12: 1

Loading program. Please wait...
PROGRAM 2-1 TO CONSTRUCT A DATA FILE

Please wait for subroutines to load.
```

Figure 1.2   Screen display of the menu of programs.

9

computers; the GW-BASIC interpreters work for the Leading Edge and Zenith computers. The compressed print option for Program 7-1 works on the Panasonic KX-P1091 printer as well as Epson printers.

If you decide to customize the programs for your particular purposes, remember that the programs are set up to use five files. Therefore, you need to specify this number of files, when you call up BASICA; on the IBM-PC, the command is **BASICA/F:5**.

You also need to familiarize yourself, if you are not familiar already, with the way in which these programs handle errors through error trapping (lines 50000 to 65010 in file SUBROUT.SUB). Because Microsoft BASIC can handle only one error at a time, if the statement **ON ERROR GOTO** directs a response into a routine in which you make a second error, this second error may cause a standard BASICA error message to appear and the program to stop.

REFERENCES

1. J. C. Hawkins, complier, 1981 *Oxford Universal Dictionary*, Oxford University Press, s.v. "data."
2. Ibid., s.v. "analyze."
3. R. J. Howarth and R. Sinding-Larsen, 1983, Multivariate Analysis: in R. J. Howarth, ed. *Handbook of Exploration Geochemistry*, Vol. 2, *Statistics and Data Analysis in Geochemical Prospecting*, Elsevier, Amsterdam, pp. 207-289.
4. D. E. Broxton, 1980, *Uranium Hydrogeochemical and Stream Sediment Reconnaissance Data Release for the Butte NTMS Quadrangle, Montana, Including Concentrations of Forty-two Additional Elements*, Los Alamos National Laboratory, Los Alamos, N.M., informal report LA-7668-MS, 207p.

CHAPTER 2

Writing and
Editing Data Files

Constructing a data file is a first step in any data analysis. In this chapter, I explain three programs:

Program 2-1 constructs a new data file.

Program 2-2 corrects mistakes in a data file.

Program 2-3 adds or deletes stations in a data file.

Program 2-1 is explained in Sections 2.1 to 2.6; Program 2-2 in Section 2.7, and Program 2-3 in Section 2.8.

2.1 OVERVIEW

Selecting Program 2-1 from the menu (Figure 1.2), you can either construct an entirely new data file or rewrite data from an existing file to the form used in the program. I also explain how to use files created by two popular software packages, WordStar and Lotus 1-2-3, with these programs.

Each data file is a random-access file structured in two parts. Part 1 consists of three lines describing the file (name, date, purpose, and so forth) and a fourth line containing the number of variables and their names. Part 2 contains the data proper, which consist of (1) from zero to five descriptive variables (station names or numbers, geologic data, lithologic data, x-coordinates, y-coordinates), and (2) from one to twenty elements or other geochemical variables.

2.2 BEGINNING TUTORIAL

Suppose you want to construct a data file for the geochemical data in Table 1.1. You can do this with Program 2-1, which I explain step by step. The program has six parts:

1. Name the data files.

2. Describe your data files.

3. Select the data-file structure.

4. Review the data-file structure.

5. How do you want to enter data? Specify whether you want to enter the data from the keyboard or from a file.

6. Enter the data for each station.

For simplicity, I first explain how to enter only two variables, the station numbers and the zinc data, for the first ten stations. Following are the step-by-step instructions. In response to the menu (Figure 1.2), enter a **1**.

Figure 2.1* is the display that appears on the screen as you key in replies to the prompts; your responses follow the colon.

This figure, and all of the other screen displays, show you everything that appears on the screen. The last line of nearly all screen displays is the highlighted message:

Press any key to continue. (The screen will clear.)

For printing in this book, this figure and the other similar ones, have been divided into blocks ending with this message. (A few screens end with another message telling you how to respond to a particular situation.)

The numbered commentary that follows will help you get started while you are learning.

1. The second line of the program asks you to wait while the program subroutines are loaded into the computer; this line appears only when you start up the system using interpreted BASIC. The next line states the time when you started the program.

2. You can identify your data file with any name that you have not used before (so that you cannot by mistake destroy a preexisting file). I entered the name **example**; DOS converts these lower-case letters to the file name **EXAMPLE**.

3. Enter three lines to describe your data file. This is the place for you to record the name, date, purpose, source of information, and so forth. The information is recorded as the first three lines of your data file. Each line can contain as many as

---

*All figures appear at the end of the chapter.

128 characters; if you wish, you can omit lines 2 and 3 or only line 3 by pressing the carriage return key.

   4. If you have described your files to your satisfaction, continue to the next step. Otherwise, by entering **N**, you return to Part 1 to re-enter Parts 1 and 2.

   5. You can include as many as five identification variables in your data file. Although the first three are ordinarily the station numbers, geologic data, and lithologic data, you can use these variables to enter other identification information. Each of these variables is from one to ten characters. The x- and y-coordinates must be numeric. For the example, I used only the first variable, which is station number. To demonstrate that the program accepts either lowercase or capital letters for "yes" (**y**), "no" (**n**), and similar responses, I used lowercase letters in this example.

   6. For this example, I selected only one variable.

   7. I entered the symbol **Zn**, for zinc. You can pick any symbols that you like, provided that all are different from one another and that each is exactly two characters (blanks count as characters).

   8. Since the data-file structure is the way I wanted it, I entered **Y**.

   9. Since I wanted to key in the data, I entered **Y**.

   10. Entering data from the keyboard simply requires that you respond to the prompts. To reduce the number of words on the screen and to make your responses line up, the prompts for stations following the first one are abbreviated. To signal that you are finished entering data, simply type **@** in place of the station number. Within Program 2-1, you do not have a chance to correct any mistakes in the file; you can change any part of the file—the header lines, variables names, or data entries—using program 2-2. You can use Program 2-3 to add more stations to your file.

   11. The last four program lines list the date, the time when the processing started, the time when processing was completed, and the time it took to construct your data file.

2.3 EXTENDED TUTORIAL A: ENTERING DATA FROM A FORMATTED DATA FILE

   Entering data from a preexisting file requires that you understand something about how files are constructed, so for now you may want to skip this section and the next unless you have data to enter from a file.

   As an example of a preexisting file, consider Table 1.2, which is a listing of a file made on a mainframe computer and transferred (downloaded) to an IBM PC. The file is arranged differently from those of this book. There are four differences:

1. Five header lines rather than three precede the data.

2. The file is formatted (each variable occupies a set number of columns) but the order of columns is different.

3. The data fields for station identification (ID), latitude, longitude, contamination code, and rock code are fewer than ten characters.

4. The file contains an end-of-record mark, which is a slash (/) on the last line (not shown on the abbreviated listing in the table).

Program 2-1 handles the third and fourth differences within the program without your intervention. Although four of the variables do not occur in the list of five descriptive variables in the system, we can replace latitude by y-coordinate, longitude by x-coordinate, contamination code by lithology, and rock code by geology. As an example of program operation, Figure 2.2 shows how to construct file XBUTTE-2 for only the station number, latitude, longitude, and zinc. The two diskettes also contain a file XBUTTE-3, which contains all the data for the thirteen stations in the original file XBUTTE.

1. After I entered the file name **XBUTTE-2**, the program checked to ensure that no file with this name existed on the disk. You will need to choose another name when you practice this procedure because, after running the program, I left the file on your disk.

2. In Part 6(B), I entered 5 to indicate the number of lines of information preceding the actual data records.

3. Unless the data are formatted they can be read only in the order of the original file. I explain the procedure in Section 2.4.

4. In response to the prompts for each variable, you need to specify three things: position in the file (index number), number of the starting column, and number of columns. This is easiest to do if you first make a table; Table 2.1 is a sample for the BUTTE data file. Then you simply enter the tabled data in response to the prompts. For instance, the starting column for station ID is one, and the number of columns for this variable is seven.

TABLE 2.1
Variable List for Data Files BUTTE and XBUTTE

Variable	Index Number	Starting Column	Number of Columns
Station ID	1	1	7
Latitude (y)	3	8	9
Longitude (x)	4	17	9
Contamination code	5	26	2
Rock code	6	28	7
Silver (Ag), ppm	7	35	7
Gold (Au), ppm	8	42	7
Copper (Cu), ppm	9	49	7
Manganese (Mn), ppm	10	56	7
Lead (Pb), ppm	11	63	7
Zinc (Zn), ppm	12	70	7

5. Part 8 allows you to review the file structure. If it is not the way you want it, the program returns you to Part 6. If it is the way you want it, the program reads in the formatted data file line by line.

2.4 EXTENDED TUTORIAL B: ENTERING DATA FROM AN UNFORMATTED DATA FILE

Table 2.2 is an example of an unformatted data file. I wrote it using WordStar's nondocument option and named it ZINC. Figure 2.3 shows how Program 2-1 creates data file ZINC-2 using this file as input. All of the responses are the same down to part 6(C) where the appropriate response is **N**. Because file ZINC-2 is on the program diskette, you will need to choose another name when you practice this procedure.

Remember, Program 2-1 does not allow you to rearrange the order of the variables in a nonformatted input file.

TABLE 2.2
WordStar File ZINC of Sample Numbers and
Zinc Values from Table 1.1

1,24
2,25
3,42
4,50
5,52
6,29
7,26
8,23
9,89
10,72
11,31
12,115
13,535
14,48
15,1010
16,560
17,48
18,44
19,36
20,33
21,45
22,118
23,274
24,81
25,80
/

2.5 EXTENDED TUTORIAL C: ENTERING DATA FOR A FILE WITHOUT STATION NAMES OR NUMBERS

Entering data for a file without station names or numbers is illustrated for file SILVER in Figure 2.4. The data are from *Statistical Analysis of Geological Data*.[1] File SILVER is not on your disk; the equivalent file FRESNI for all stations in the reference is used in Section 2.7 to show you how to correct a data file.

2.6 MORE DETAILS ON PREEXISTING FILES

You have learned that Program 2-1 will accept data from the keyboard or from a file. If a file is used, it must be an ASCII file that may be downloaded from a mainframe computer, written using WordStar, written with Lotus 1-2-3, or written in another way. The file may be formatted or unformatted. If unformatted, the variables must be separated by commas.

For the first two stations, I prepared identically formatted files using WordStar and Lotus 1-2-3 (Table 2.3). If using WordStar, you need to remember to write a nondocument file; this option produces an ASCII file. Using Lotus 1-2-3, you need to save a file as a print file, unformatted (so that the three blank lines and the header lines normally provided by the program are not produced), with the left margin set to 0 and the right margin set to a number as wide as or wider than the number of columns in the file. You can process this file as you did the previous ones.

2.7 CORRECTING STATIONS IN A DATA FILE

You often will want to correct stations in a data file either because you keyed in the data incorrectly or because there were mistakes in the original data list. Program 2-2 enables you to do this. The program has four parts:

1. Name the data file to correct.

2. File description.

3. File structure.

4. Edit records.

Following are step-by-step instructions for changing the data file EXAMPL-2. This file is a duplicate of file EXAMPLE, created with the beginning tutorial in Section 2.2, but not provided on your diskette. On your diskette, file EXAMPL-2 is the original one, before the changes explained in this section were made.

Figure 2.5 is the display that appears on the screen as you key in replies to the prompts.

16

TABLE 2.3
Listing of Data File YBUTTE Written with Lotus 1-2-3

GEOCHEMICAL DATA FROM THE BUTTE QUADRANGLE BETWEEN 112 AND 113 WEST AND
46 AND 46.25 NORTH.. 265 LOCATIONS. THE FORMAT IS A7,2F9.4,I2,A7,7I7.
VARIABLES ARE SAMPLE NUMBER, LATITUDE, LONGITUDE, CONTAMINATION CODE,
ROCK CODE, AG, AU, CU, MN, PB, AND ZN.

| M24164 | 46.0539 | 112.6022 | 3 | T-CV-- | 2 | 2 | 32 | 659 | 9 | 95 |
| M25319 | 46.0144 | 112.9803 | 0 | T-CV-- | 2 | 2 | 94 | 554 | 52 | 96 |

17

1. Enter the name of the data file that you want to correct. I entered **EXAMPL-2**.

2. You can change one or more of the header lines; I changed lines 1 and 2.

3. Part 3(C) allows you to change the name of one or more elements; I kept the name Zn the same.

4. After you enter the station name or number you want to correct, the program searches the file to find a match. Then it prints the station identification and the values of other variables.

5. Now you can change one or more items. To illustrate, I changed the erroneous value of 52 to 62 for zinc in station 5. If you create a data file without station numbers, Program 2-1 assigns them sequentially starting with number 1. To edit the file, you simply need to enter the appropriate sequence number. Figure 2.6 shows you how to do this for data file FRESNI, constructed in Section 2.5. By entering **15** in response to the prompt for station number, you get the data for the 15th station entered. Because the entry was correct, I of course did not change it.

2.8 ADDING OR DELETING STATIONS IN A DATA FILE

Besides correcting stations in a data file, as explained in Section 2.6, you may want to add or delete stations from a file. Program 2-3 does this. The program has these six parts:

1. Name the data file.

2. File description.

3. File structure.

4. Choose to either add or delete stations.

5. Depending on your selection in Part 4, add or delete stations.

6. If you chose to add stations in Part 5, the program gives you a chance to delete stations, or, if you chose to delete stations in Part 5, you now have a chance to add one or more here.

Here are the step-by-step instructions for changing your data file EXAMPL-3, which is an exact copy of the file constructed in Section 2.2.

Figure 2.7 is the display that appears on the screen while you run this program.

1. Enter the name of the data file in which you want to add or delete stations. I entered **EXAMPL-3**.

2. In Part 4, you type **A** to add a station, or **D** to delete one. If you want both to add and delete stations, you can start with either option, since the program will always lead you to the other option.

3. To illustrate, I show you how to delete station 2.

4. Because station 2 has been deleted from the file, it does not appear when I select it a second time.

5. Here is an illustration of adding another station, say number 11. Note that the program asks you to enter information for station number 10, because there are only nine stations rather than the ten listed in Part 3, one having been deleted in Part 5. You might need to add this station because you had stopped part way through keying in the data list from Table 1.1 and were now resuming work.

REFERENCE

1. G. S. Koch, Jr. and R. F. Link, 1980, *Statistical Analysis of Geological Data*, Vol. 2, Dover Publications, New York, p. 9. (Reprinted, with corrections, from the original two volumes published in 1971-1972 by Wiley, New York.)

Figure 2.1 Screen display of constructing data file EXAMPLE.

```
PROGRAM 2-1 TO CONSTRUCT A DATA FILE

Please wait for subroutines to load.

Processing begun at 15:27:39 on 08-19-1986

Press any key to continue.  (The screen will clear.)

PART 1.  NAME THE DATA FILE

(A) Enter the name of the data file: example

Press any key to continue.  (The screen will clear.)

PART 2.  DESCRIBE YOUR DATA FILE

(A) Three lines have been left to enter
    information that describes your data file:

Enter line 1: Example of file construction
Enter line 2: Zinc data from table 1.1
Enter line 3: 19 August 1986
```

Data file: example

(B) If it's the way you want it, enter 'Y'; if not, enter 'N': Y

Press any key to continue. (The screen will clear.)

PART 3.   SELECT THE DATA-FILE STRUCTURE

(A) Do you want to include the following variables?

Do you want to include Station ID (Y/N): y
Do you want to include Geology   (Y/N): n
Do you want to include Lithology   (Y/N): n
Do you want to include X-coord.   (Y/N): n
Do you want to include Y-coord.   (Y/N): n

(B) How many elements or other geochemical variables do you have? (1 to 20): 1

(C) Choose symbols for this (these) variable(s):

    Enter a 2-character symbol for variable number 1: Zn

Press any key to continue. (The screen will clear.)

(continued)

Figure 2.1 *(continued)*

PART 4. REVIEW THE DATA-FILE STRUCTURE.

(A) The file contains this (these) identification
     and location variable(s):

Station ID

(B) The file contains this (these) chemical element(s):

Zn

(C) No stations

(D) If it's the way you want it, enter 'Y'; if not, enter 'N': Y

     Press any key to continue. (The screen will clear.)

PART 5. HOW DO YOU WANT TO ENTER DATA

Do you want to key in data? (Y/N): y

     Press any key to continue. (The screen will clear.)

PART 6. ENTER DATA FOR EACH STATION

Press @ when you have entered all your data

For station number 1
Enter data for Station ID (Up to 10 characters): 1
Enter data for Zn: 24

For station number 2
Station ID: 2
Zn: 250

For station number 3
Station ID: 3
Zn: 42

For station number 4
Station ID: 4
Zn: 50

(continued)

Figure 2.1 *(continued)*

```
For station number    5
Station ID: 5
Zn: 52

For station number    6
Station ID: 6
Zn: 29

For station number    7
Station ID: 7
Zn: 26

For station number    8
Station ID: 8
Zn: 23

For station number    9
Station ID: 9
Zn: 89
```

For station number   10
Station ID: 10
Zn: 72

For station number   11
Station ID: @

Press any key to continue. (The screen will clear.)

Date: 08-19-1986
Processing started   at 15:27
Processing completed at 15:33
Elapsed time   0 hour(s), 6 minute(s)

Figure 2.2 Screen display of constructing data file XBUTTE-2 from preexisting data file XBUTTE.

```
PROGRAM 2-1 TO CONSTRUCT A DATA FILE

Processing begun at 08:47:37 on 08-20-1986

Press any key to continue. (The screen will clear.)

PART 1.  NAME THE DATA FILE

(A) Enter the name of the data file: XBUTTE-2

Press any key to continue. (The screen will clear.)

PART 2.  DESCRIBE YOUR DATA FILE

(A) Three lines have been left to enter
    information that describes your data file:

Enter line 1: Example of data-file construction
Enter line 2: Zinc data from table 1.2
Enter line 3: 20 August 1986
```

Data file: XBUTTE-2

(B) If it's the way you want it, enter 'Y'; if not, enter 'N': Y

Press any key to continue. (The screen will clear.)

PART 3.   SELECT THE DATA-FILE STRUCTURE

(A) Do you want to include the following variables?

Do you want to include Station ID   (Y/N): Y
Do you want to include Geology   (Y/N): N
Do you want to include Lithology   (Y/N): N
Do you want to include X-coord.   (Y/N): Y
Do you want to include Y-coord.   (Y/N): Y

(B) How many elements or other geochemical variables do you have? (1 to 20): 1

(C) Choose symbols for this (these) variable(s):

    Enter a 2-character symbol for variable number 1: Zn

Press any key to continue. (The screen will clear.)

(continued)

Figure 2.2 *(continued)*

PART 4. REVIEW THE DATA-FILE STRUCTURE.

(A) The file contains this (these) identification
    and location variable(s):

Station ID
X-coord.
Y-coord.

(B) The file contains this (these) chemical element(s):

Zn

(C) No stations

(D) If it's the way you want it, enter 'Y'; if not, enter 'N': Y

Press any key to continue. (The screen will clear.)

PART 5. HOW DO YOU WANT TO ENTER DATA

Do you want to key in data? (Y/N): N

Press any key to continue. (The screen will clear.)

PART 6. SPECIFY YOUR ORIGINAL DATA FILE STRUCTURE AND NAME

(A) Enter the name of your sequential file: XBUTTE

(B) How many lines of identification information
     precede the first data record?: 5

(C) Are the data in your file formatted? (Y/N): ? Y

Press any key to continue. (The screen will clear.)

PART 7. SPECIFY FILE FORMAT

Enter the starting column for Station ID: 1
Enter the number of columns for Station ID: 7

(continued)

Figure 2.2 *(continued)*

Enter the starting column for X-coord.: 17
Enter the number of columns for X-coord.: 9

Enter the starting column for Y-coord.: 8
Enter the number of columns for Y-coord.: 9

Enter the starting column for Zn: 70
Enter the number of columns for Zn: 7

PART 8.   REVIEW THE INPUT DATA-FILE STRUCTURE.

(A) File name: XBUTTE
(B) Number of identification lines: 5
(C) Variable name, starting column, number of columns

Station ID, 1 , 7
  X-coord., 17 , 9
  Y-coord., 8 , 9
Zn, 70 , 7

(D) If it's the way you want it, enter 'Y'; if not, enter 'N': Y

Press any key to continue. (The screen will clear.)

Now reading line   1   2   3   4   5   6   7   8   9   10   11   12
13

Press any key to continue. (The screen will clear.)

Figure 2.3   Screen display of constructing data file ZINC-2 from preexisting data file ZINC.

```
PROGRAM 2-1 TO CONSTRUCT A DATA FILE

Processing begun at 09:00:33 on 08-20-1986

Press any key to continue. (The screen will clear.)

PART 1.   NAME THE DATA FILE

(A) Enter the name of the data file: ZINC-2

Press any key to continue. (The screen will clear.)

PART 2.   DESCRIBE YOUR DATA FILE

(A) Three lines have been left to enter
    information that describes your data file:

Enter line 1: Third example of file construction
Enter line 2: Zinc data from table 2.2
Enter line 3: 20 August 1986
```

Data file: ZINC-2

(B) If it's the way you want it, enter 'Y'; if not, enter 'N': Y

Press any key to continue. (The screen will clear.)

PART 3. SELECT THE DATA-FILE STRUCTURE

(A) Do you want to include the following variables?

Do you want to include Station ID (Y/N): Y
Do you want to include Geology   (Y/N): N
Do you want to include Lithology   (Y/N): N
Do you want to include X-coord.   (Y/N): N
Do you want to include Y-coord.   (Y/N): N

(B) How many elements or other geochemical variables do you have? (1 to 20): 1

(C) Choose symbols for this (these) variable(s):

   Enter a 2-character symbol for variable number 1: Zn

Press any key to continue. (The screen will clear.)

(continued)

Figure 2.3 *(continued)*

PART 4.   REVIEW THE DATA-FILE STRUCTURE.

(A) The file contains this (these) identification
    and location variable(s):

Station ID

(B) The file contains this (these) chemical element(s):

Zn

(C) No stations

(D) If it's the way you want it, enter 'Y'; if not, enter 'N': Y

    Press any key to continue. (The screen will clear.)

PART 5. HOW DO YOU WANT TO ENTER DATA

Do you want to key in data? (Y/N): N

    Press any key to continue. (The screen will clear.)

PART 6. SPECIFY YOUR ORIGINAL DATA FILE STRUCTURE AND NAME

(A) Enter the name of your sequential file: ZINC

(B) How many lines of identification information
    precede the first data record?: 4

(C) Are the data in your file formatted? (Y/N): ? N

Press any key to continue. (The screen will clear.)

Now reading line   1   2   3   4   5   6   7   8   9   10   11   12
13   14   15   16   17   18   19   20   21   22   23   24
Press any key to continue. (The screen will clear.)

Figure 2.4 Screen display of constructing data file SILVER without station names or numbers.

PROGRAM 2-1 TO CONSTRUCT A DATA FILE

Processing begun at 09:07:19 on 08-20-1986

Press any key to continue. (The screen will clear.)

PART 1. NAME THE DATA FILE

(A) Enter the name of the data file: SILVER

Press any key to continue. (The screen will clear.)

PART 2. DESCRIBE YOUR DATA FILE

(A) Three lines have been left to enter
    information that describes your data file:

Enter line 1: Observations of silver content, Fresnillo mine, Mexico
Enter line 2: From Statistical Analysis of Geological Data, v. 2, p. 9
Enter line 3: 20 August 1986

Data file: SILVER

(B) If it's the way you want it, enter 'Y'; if not, enter 'N': Y

Press any key to continue. (The screen will clear.)

PART 3. SELECT THE DATA-FILE STRUCTURE

(A) Do you want to include the following variables?

Do you want to include Station ID (Y/N): N
Do you want to include Geology   (Y/N): N
Do you want to include Lithology (Y/N): N
Do you want to include X-coord.   (Y/N): Y
Do you want to include Y-coord.   (Y/N): N

(B) How many elements or other geochemical variables do you have? (1 to 20): 1

(C) Choose symbols for this (these) variable(s):

    Enter a 2-character symbol for variable number 1: Ag

Press any key to continue. (The screen will clear.)

(continued)

Figure 2.4 *(continued)*

PART 4. REVIEW THE DATA-FILE STRUCTURE.

(A) The file contains this (these) identification
    and location variable(s):

    X-coord.

(B) The file contains this (these) chemical element(s):

Ag

(C) No stations

(D) If it's the way you want it, enter 'Y'; if not, enter 'N': Y

    Press any key to continue. (The screen will clear.)

PART 5. HOW DO YOU WANT TO ENTER DATA

Do you want to key in data? (Y/N): Y

    Press any key to continue. (The screen will clear.)

PART 6.   ENTER DATA FOR EACH STATION

Press @ when you have entered all your data

For station number 1
Enter data for X-coord.: 2
Enter data for Ag: 698

For station number 2
    X-coord.: 4
Ag: 365

For station number 3
    X-coord.: 6
Ag: 223

For station number 4
    X-coord.: 8
Ag: 335

For station number 5
    X-coord.: 10
Ag: 156

For station number 6
    X-coord.: @

Press any key to continue. (The screen will clear.)

Figure 2.5 Screen display for correcting stations in data file EXAMPLE-2.

```
PROGRAM 2-2 TO CORRECT STATIONS IN A DATA FILE

Please wait for subroutines to load.

Processing begun at 09:58:06 on 08-25-1986

Press any key to continue. (The screen will clear.)

PART 1. NAME THE DATA FILE

(A) Enter the name of the data file: EXAMPL-2

Press any key to continue. (The screen will clear.)
```

PART 2. FILE DESCRIPTION

Header lines

Example of file construction
Zinc data from table 1.1
19 August 1986

Do you want to change the header lines? (Y/N): Y
Enter line 1: Example of file construction
Enter line 2: Revised zinc data from table 1.1
Enter line 3: 25 August 1986

Press any key to continue. (The screen will clear.)

(continued)

Figure 2.5 (continued)

PART 3. FILE STRUCTURE

(A) The file contains this (these) identification
    and location variable(s):

Station ID

(B) The file contains this (these) chemical element(s):

Zn

(C) The file contains 10 station(s).

Do you want to change the element name(s)? (Y/N): N
Press any key to continue. (The screen will clear.)

Press <ENTER> to keep old value. Press @ to quit.

PART 4. EDIT RECORDS

Enter station name or number to edit: 5

Station ID: 5
Zn: 52

Do you want to change any item(s)? (Y/N): Y

If you want data to remain the same, press only the <CR> key.
Otherwise, enter the changed item.

Enter data for Station ID:
Enter data for Zn: 62

Enter station name or number to edit: @

Press any key to continue. (The screen will clear.)

Figure 2.6 Screen display for correcting stations in data file FRESNI.

```
PROGRAM 2-2 TO CORRECT STATIONS IN A DATA FILE

Processing begun at 11:38:12 on 08-25-1986


Press any key to continue. (The screen will clear.)

PART 1. NAME THE DATA FILE

(A) Enter the name of the data file: FRESNI

Press any key to continue. (The screen will clear.)

PART 2. FILE DESCRIPTION

Header lines

Data file FRESNI
From Koch and Link, Statistical Analysis, 1971, p. 9
22 August 1986

Do you want to change the header lines? (Y/N): N
Press any key to continue. (The screen will clear.)
```

PART 3. FILE STRUCTURE

(A) The file contains this (these) identification
    and location variable(s):

    X-coord.

(B) The file contains this (these) chemical element(s):

Ag

(C) The file contains 22 station(s).

Do you want to change the element name(s)? (Y/N): N
Press any key to continue. (The screen will clear.)

Press <ENTER> to keep old value. Press @ to quit.

PART 4. EDIT RECORDS

Enter station name or number to edit: 15

X-coord.:   30.0000
Ag: 65

Do you want to charge any item(s)? (Y/N): N

Enter station name or number to edit: @

Press any key to continue. (The screen will clear.)

46

Figure 2.7 Screen display for adding or deleting stations in data file EXAMPLE-2.

```
PROGRAM 2-3 TO ADD OR DELETE STATIONS IN A DATA FILE

Please wait for subroutines to load.

Processing begun at 16:02:41 on 08-26-1986

Press any key to continue. (The screen will clear.)

PART 1. NAME THE DATA FILE

(A) Enter the name of the data file: EXAMPL-2

Press any key to continue. (The screen will clear.)

PART 2. FILE DESCRIPTION

Header lines

Example of file construction
Zinc data from table 1.1
19 August 1986
```

Press any key to continue. (The screen will clear.)

PART 3. FILE STRUCTURE

(A) The file contains this (these) identification
    and location variable(s):

Station ID

(B) The file contains this (these) chemical element(s):

Zn

(C) The file contains 10 station(s).
    Press any key to continue. (The screen will clear.)

Press @ when you have entered all your data

PART 4. CHOOSE TO ADD OR DELETE STATIONS

Do you want to ADD or DELETE a station? (A/D): D
    Press any key to continue. (The screen will clear.)

Press @ when you have entered all your data

(continued)

Figure 2.7 *(continued)*

```
PART 5.  DELETE STATIONS

Enter station name or number to DELETE: 2

Station ID: 2
Zn: 25
Do you want to delete this station? (Y/N): Y

Enter station name or number to DELETE: 2
Station not found

Enter station name or number to DELETE: @

   Press any key to continue. (The screen will clear.)
```

PART 6. CHOOSE TO ADD STATIONS

Do you want to ADD stations? (Y/N): Y
    Press any key to continue. (The screen will clear.)

Press @ when you have entered all your data
PART 5. ADD STATIONS

For Station number 10
Enter data for Station ID: 11
Enter data for Zn: 35

For Station number 11
Enter data for Station ID: @

    Press any key to continue. (The screen will clear.)

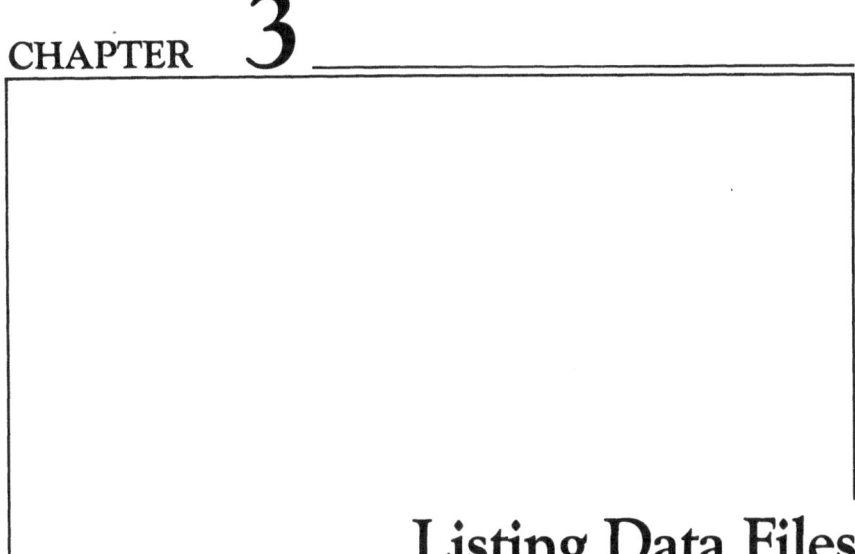

CHAPTER **3**

Listing Data Files

You will want to list data files either for your use or for presentation to others. The first lists will be for your editing, so that you can verify that the files prepared with the programs in Chapter 1 are complete and accurate. Once the files are correct, you will want lists for records. Scanning the lists will detect patterns in the data that are obvious from visual inspection. Generally, your lists will include all of the identification and chemical variables.

For presentation to others, you may want partial lists, perhaps limited to the most interesting variables or to those that will fit in the width of a single page.

3.1 OVERVIEW

Program 3-1 creates lists that can either be printed or displayed on the screen for all or selected variables in a data file. You can print these lists using either normal type (10 characters per inch in a horizontal direction) or compressed type (17.16 characters per inch). You can start each page of a listing with as many as ten title lines.

3.2 BEGINNING TUTORIAL

Using program 3-1, you can list file EXAMPLE, which I made following the instructions in Section 2.3. Program 3-1 has seven parts:

51

1. Name the data file.

2. Specify the output device (screen or printer).

3. File description.

4. File structure.

5. Specify the list title.

6. Select variables for the list.

7. Specify the output format.

Following are the step-by-step instructions (Figure 3.1*):

1. You may choose screen output if you want simply to determine that your data file contains the variables you want or is free of input errors. Screen output is paged with the message **"Press any key to continue"** printed at the bottom of each page. If you select printer output, the next question asks you whether you want normal or compressed type.

2. The listing is that of the original file EXAMPLE constructed in Section 2.2. Of course, if you changed this file in subsequent practice sessions, your list will be different from this output. Alternatively, you can copy the original file EXAMPLE from your backup disk to your working diskette.

3. Since there is only one variable in the file, there is no question about what to list here.

3.3 EXTENDED TUTORIAL A: LISTING DATA FILE XBUTTE-3

File XBUTTE-3 contains full data for the 13 stations listed in Table 1.2 of the 265 stations in data file BUTTE. In this section, you will learn how to list this data file on the printer by keying in the responses of Figure 3-2. Following are notes for your work:

1. Here, you answer **Y**, in order to get output on the printer.
2. Using compressed type, all six chemical variables will list on the print line.

3.4 EXTENDED TUTORIAL B: LISTING DATA FILES BUTTE-2 AND NORWAY-2

You can follow the procedure in Section 3.3 to get listings for data files BUTTE-2 and NORWAY-2 The listing for BUTTE-2 took 16 minutes to print on an Epson Spectrum LX-80 printer.

*All figures appear at the end of the chapter.

3.5 SPECIFYING THE OUTPUT FORMAT

In Part 7, you are asked to specify output formats with from 0 to 5 decimal places; the program allocates 6 characters for each column. I choose a 6-character-wide column as one wide enough to display most data and get several columns on one page. If you specify too many decimal places for a particular value, the program prints the value preceded by a percent sign (%) to indicate your error. For example, if you try to print the first value for Mn in file XBUTTE-3 using three decimal points, the program will print **%659.000**. If you know the largest value of each variable in a file, you can specify the decimal point position without trouble. If you do not you may need to run the program twice: once to find the largest values, and a second time for decimal point placement.

I did not provide for the program to place the decimal points automatically because doing so would have required reading through each data list twice.

Figure 3.1   Screen display for listing data file EXAMPLE.

```
PROGRAM 3-1 TO LIST A DATA FILE

Processing begun at 11:32:37 on 08-20-1986

Press any key to continue. (The screen will clear.)

PART 1. NAME THE DATA FILE

(A) Enter the name of the data file: EXAMPLE

Press any key to continue. (The screen will clear.)

PART 2. SPECIFY THE OUTPUT DEVICE

(A) Do you want the output routed to printer? (Y/N): N

Press any key to continue. (The screen will clear.)
```

PART 3. DESCRIPTION OF THE DATA FILE

Header lines

Example of file construction
Zinc data from table 1.1
19 August 1986

Press any key to continue. (The screen will clear.)

PART 4. STRUCTURE OF THE DATA FILE

(A) The file contains this (these) identification
    and location variable(s):

Station ID

(B) The file contains this (these) chemical element(s):

Zn

(C) The file contains 10 station(s).

The description of the data file on the screen will disappear.
Write down any of it that you will need.

Press any key to continue. (The screen will clear.)

(continued)

Figure 3.1 *(continued)*

PART 5. SPECIFY THE LIST TITLE

(A) How many title lines do you want? Enter a number between 0 and 10 : 3

Enter each title line (0 to 79 characters) in response to the prompt(s)

    Enter line 1: Listing of file EXAMPLE
    Enter line 2: Zinc data from Table 1.1
    Enter line 3: 20 August 1986

Press any key to continue. (The screen will clear.)

PART 6. SELECT VARIABLES FOR THE LIST

Do you want to list the Station ID (Y/N): Y

(A) How many of the 1 variables do you want to list? 1

Working from the leftmost column,

    Enter the number of the variable (1- 1) for column 1: 1

Press any key to continue. (The screen will clear.)

PART 7. SPECIFY THE OUTPUT FORMAT

Enter units for Zn (6 or fewer characters): ppm
Enter the number of decimal places (0-5): 0

Press any key to continue. (The screen will clear.)

Date: 08-20-1986 Time: 11:35:12   Page 1

Listing of file EXAMPLE
Zinc data from Table 1.1
20 August 1986

Station ID	Zn ppm
1	24
2	25
3	42
4	50
5	52
6	29
7	26
8	23
9	89
10	72

Press any key to continue. (The screen will clear.)

Figure 3.2   Screen display for listing data file XBUTTE-3.

```
RUN
PROGRAM 3-1 TO LIST A DATA FILE

Processing begun at 11:57:41 on 08-20-1986

Press any key to continue. (The screen will clear.)

PART 1. NAME THE DATA FILE

(A) Enter the name of the data file: XBUTTE-3

Press any key to continue. (The screen will clear.)

PART 2. SPECIFY THE OUTPUT DEVICE

(A) Do you want the output routed to printer? (Y/N): Y

    Turn on the printer and set the paper.
```

(B) Print with compressed type? (Y/N) : Y

Press any key to continue. (The screen will clear.)

PART 3. DESCRIPTION OF THE DATA FILE

Header lines

File XBUTTE-3
Data from Table 1.2
20 August 1986

Press any key to continue. (The screen will clear.)

PART 4. STRUCTURE OF THE DATA FILE

(A) The file contains this (these) identification
    and location variable(s):

Station ID
Geology
Lithology
    X-coord.
    Y-coord.

(continued)

Figure 3.2 (continued)

(B) The file contains this (these) chemical element(s):

Ag
Au
Cu
Mn
Pb
Zn

(C) The file contains 13 station(s).

The description of the data file on the screen will disappear.
Write down any of it that you will need.

Press any key to continue. (The screen will clear.)

PART 5. SPECIFY THE LIST TITLE

(A) How many title lines do you want? Enter a number between 0 and 10): 3

Enter each title line (0 to 79 characters) in response to the prompt(s)

    Enter line   1: Butte quadrangle data from Table 1.2
    Enter line   2: Full data listing
    Enter line   3: 20 August 1986

Press any key to continue. (The screen will clear.)

PART 6.   SELECT VARIABLES FOR THE LIST

Do you want to list the Station ID (Y/N): Y
Do you want to list the Geology   (Y/N): Y
Do you want to list the Lithology (Y/N): Y
Do you want to list the   X-coord. (Y/N): Y
Do you want to list the   Y-coord. (Y/N): Y

(A) How many of the 6 variables do you want to list? 6

Working from the leftmost column,

    Enter the number of the variable (1- 6) for column 1: 1

(continued)

Figure 3.2 *(continued)*

```
Enter the number of the variable (1- 6 ) for column  2: 2

Enter the number of the variable (1- 6 ) for column  3: 3

Enter the number of the variable (1- 6 ) for column  4: 4

Enter the number of the variable (1- 6 ) for column  5: 5

Enter the number of the variable (1- 6 ) for column  6: 6

Press any key to continue. (The screen will clear.)

PART 7. SPECIFY THE OUTPUT FORMAT

Enter units for Ag (6 or fewer characters): ppm
Enter the number of decimal places (0-5): 0
```

Enter units for Au (6 or fewer characters): ppm
Enter the number of decimal places (0-5): 0

Enter units for Cu (6 or fewer characters): ppm
Enter the number of decimal places (0-5): 0

Enter units for Mn (6 or fewer characters): ppm
Enter the number of decimal places (0-5): 0

Enter units for Pb (6 or fewer characters): ppm
Enter the number of decimal places (0-5): 0

Enter units for Zn (6 or fewer characters): ppm
Enter the number of decimal places (0-5): 0

Press any key to continue. (The screen will clear.)

(continued)

Figure 3.2 *(continued)*

Date: 08-20-1986   Time: 12:07:09   Page 1

Butte quadrangle data from Table 1.2
Full data listing
20 August 1986

Station ID	Geology	Lithology	X-coord.	Y-coord.	Ag ppm	Au ppm	Cu ppm	Mn ppm	Pb ppm	Zn ppm
M24164	T-CV--	3	112.6022	46.0539	2	2	32	659	9	95
M25319	T-CV--	0	112.9803	46.0144	2	2	94	554	52	96
M24449	T-CV--	9	112.5694	46.2172	2	2	31	930	41	85
M24196	T-CV--	3	112.5139	46.0931	14	2	68	1517	12	101
M24459	T-CV--	3	112.5250	46.2236	2	2	33	1746	12	34
M24114	T-CV--	3	112.5597	46.1833	2	2	76	427	31	89
M25341	T-CV--	0	112.8114	46.0158	2	2	63	1235	20	71
M24178	T-CV--	0	112.5844	46.1317	2	2	90	584	40	93
M24090	T-CV--	8	112.5403.	46.1569	2	2	62	634	28	128
M25313	T-CV--	0	112.9339	46.1589	2	2	821	496	83	153
M24211	T-CV--	3	112.6708	46.0408	2	2	62	495	36	143
M24117	T-CV--	9	112.5625	46.2194	2	2	79	737	44	138
M24170	T-CV--	3	112.5975	46.0806	2	10	39	333	23	69

CHAPTER 4

Summary Statistics

If a data file contains data for more than a few stations, it is impossible to make sense out of it without calculating statistics to summarize the information. Even if there are only a few stations, summary statistics—for example, the mean, standard deviation, and confidence interval for the mean—will let you compare the data set with another one.

4.1 OVERVIEW

Program 4-1 provides two kinds of tables of summary statistics. The first is an abbreviated table containing statistics for all the variables that you select from your data list. The second is a detailed table for each variable that you select from the first table.

For each specified variable, the first table lists the symbol, the number of samples (total number and the number above the analytical detection limit), the mean, the upper and lower confidence limits at your selected risk level, the standard deviation, and the coefficient of variation.

The second table is produced for each of the variables in the first table for which you require detailed output. The output in addition to that in the first table is the variance, confidence limits for the variance, the minimum and maximum value of the variable, and a frequency distribution.

4.2 BEGINNING TUTORIAL

The beginning tutorial explains how to calculate an abbreviated table for all variables and a detailed table for the zinc data in the NORWAY-2 data file (Table 1.1).

Program 4-1 has seven parts:

1. Name the data file.

2. File description.

3. File structure.

4. Select variables.

5. Specify parameters for the frequency distribution.

6. Specify the output device.

7. Specify detailed printer output (if required).

Following are the step-by-step instructions for the sample run (Figure 4.1*):

1. Parts 1 through 3 follow the same form as for previous programs.

2. In Part 4, you select the variables to appear in the first table and specify for each of them the number of decimal places. (The comments about decimal places in Section 3.5 also apply here.) I selected all six variables in the file.

3. In Part 5, you first specify the percentage risk level, typically 5 or 10 percent (must be a positive value). You then specify parameters for the frequency distribution of each of the variables that you selected in Part 4. The starting values and detection limits can be negative or zero values; the class intervals must, of course, be positive.

Even if you later decide not to print these frequency distributions, you need to specify these parameters. Otherwise, the program would need to reread the data list. Specifying a detection limit allows you to ignore values below it if you wish. I illustrate the use of this option in the extended tutorial (Section 4.3).

4. The results appear in the table. The means and standard deviations agree with those in Howarth and Sinding-Larsen except for the mean of Fe, which is evidently printed incorrectly in the reference.[1]

5. In Part 7, you specify **Zn** as a variable for detailed output, which always appears on the printer, since it will take more than one screen.

6. Finally, you enter @ to indicate that you are finished making detailed tables.

---

*All figures appear at the end of the chapter.

4.3 EXTENDED TUTORIAL

Figure 4.2 illustrates the use of Program 4-1 to obtain summary statistics and a frequency distribution of silver values above the detection limit of 2 ppm in data file BUTTE-2.

REFERENCE

1. R. J. Howarth and R. Sinding-Larsen, 1983, Multivariate Analysis, in *Handbook of Exploration Geochemistry*, Vol. 2, *Statistics and Data Analysis in Geochemical Prospecting*, ed. R. J. Howarth, Elsevier, Amsterdam, p. 212, Table 6-III.

Figure 4.1 Screen display of summary statistics for data file NORWAY-2.

```
PROGRAM 4-1 SUMMARY STATISTICS

Processing begun at 08:30:07 on 08-21-1986

Press any key to continue. (The screen will clear.)

PART 1. NAME THE DATA FILE

(A) Enter the name of the data file: NORWAY-2

Press any key to continue. (The screen will clear.)

PART 2. FILE DESCRIPTION

Header lines

File NORWAY-2
Constructed from Table 1.1
20 August 1986

Press any key to continue. (The screen will clear.)
```

PART 3. FILE STRUCTURE

(A) The file contains this (these) identification
    and location variable(s):

Station ID

(B) The file contains this (these) chemical element(s):

Zn
Fe
Mn
Cd
Cu
Pb

(C) The file contains 25 station(s).
    Press any key to continue. (The screen will clear.)

PART 4. SELECT VARIABLES

Do you want a tabulation for Zn? (Y/N): Y
Enter the number of decimal places for output (0-3): 2

(continued)

Figure 4.1 (continued)

Do you want a tabulation for Fe? (Y/N): Y
Enter the number of decimal places for output (0-3): 2

Do you want a tabulation for Mn? (Y/N): Y
Enter the number of decimal places for output (0-3): 2

Do you want a tabulation for Cd? (Y/N): Y
Enter the number of decimal places for output (0-3): 2

Do you want a tabulation for Cu? (Y/N): Y
Enter the number of decimal places for output (0-3): 2

Do you want a tabulation for Pb? (Y/N): Y
Enter the number of decimal places for output (0-3): 2

Press any key to continue. (The screen will clear.)

PART 5. SPECIFY PARAMETERS FOR THE FREQUENCY DISTRIBUTION

Enter the percentage risk level: 10

For Zn:
a) Enter the starting value: 0
b) Enter the class interval for the frequency distribution: 50
c) Enter the detection limit: 0

For Fe:
a) Enter the starting value: 0
b) Enter the class interval for the frequency distribution: 1
c) Enter the detection limit: 0

For Mn:
a) Enter the starting value: 0
b) Enter the class interval for the frequency distribution: 100
c) Enter the detection limit: 0

For Cd:
a) Enter the starting value: 0
b) Enter the class interval for the frequency distribution: .5
c) Enter the detection limit: 0

For Cu:
a) Enter the starting value: 0
b) Enter the class interval for the frequency distribution: 1
c) Enter the detection limit: 0

(continued)

Figure 4.1 *(continued)*

For Pb:
a) Enter the starting value: 0
b) Enter the class interval for the frequency distribution: 1
c) Enter the detection limit: 0

Enter 'A' for an ascending, or 'D' for a descending
relative-cumulative-frequency distribution: A

Calculating. Please wait...

Press any key to continue. (The screen will clear.)

PART 6. SPECIFY THE OUTPUT DEVICE

(A) Do you want the output routed to printer? (Y/N): N

Press any key to continue. (The screen will clear.)

File NORWAY-2
Constructed from Table 1.1
20 August 1986

Vari-able	No. of Samples		Mean	Confidence Limit		Standard Deviation	Coef. of Variation
	Total	>DL*		Lower	Upper		
Zn	25	25	139.60	60.48	218.72	231.22	1.66
Fe	25	25	2.26	1.76	2.75	1.44	0.64
Mn	25	25	871.20	492.27	1250.13	1107.40	1.27
Cd	25	25	1.13	0.69	1.58	1.30	1.15
Cu	25	25	73.60	20.10	127.10	156.35	2.12
Pb	25	25	9.72	8.09	11.35	4.77	0.49

* > DL = Greater than detection limit

Press any key to continue. (The screen will clear.)

PART 7. DETAILED PRINTER OUTPUT

Do you want detailed output to the printer (Y/N): Y

(continued)

Figure 4.1 *(continued)*

Turn on the printer and set the paper.

Press any key when ready.

Press @ when you have entered all your data

For detailed output to the printer,
Enter the name of the desired variable: Zn

File NORWAY-2
Constructed from Table 1.1
20 August 1986

+++VARIABLE: Zn

Total observations = 25   Number above detection limit = 25

T(10)% = 1.71

Mean	Variance	Standard Deviation	Coefficient of Variation
139.60	53464.25	231.22	1.66

Confidence Limits	Lower	Upper
Mean	60.48	218.72
Variance	35233.42	92667.74

Minimum value: 23.00
Maximum value: 1010.00

FREQUENCY DISTRIBUTION

25 values from 25 observations
Class interval = 50

Number	Class Interval		Frequency	Relative Freq.(%)	Cumulative Frequency	Relative Cum. Freq. (%)
1	0.00	- 50.00	2	8.00	2	8.00
2	50.00	- 100.00	14	56.00	16	64.00
3	100.00	- 150.00	5	20.00	21	84.00
4	150.00	- 200.00	0	0.00	21	84.00
5	200.00	- 250.00	0	0.00	21	84.00
6	250.00	- 300.00	1	4.00	22	88.00
7	300.00	- 350.00	0	0.00	22	88.00
8	350.00	- 400.00	0	0.00	22	88.00
9	400.00	- 450.00	0	0.00	22	88.00
10	450.00	- 500.00	0	0.00	22	88.00
11	500.00	- 550.00	0	0.00	22	88.00
12	550.00	- 600.00	2	8.00	24	96.00
13	600.00	- 650.00	0	0.00	24	96.00
14	650.00	- 700.00	0	0.00	24	96.00
15	700.00	- 750.00	0	0.00	24	96.00
16	750.00	- 800.00	0	0.00	24	96.00
17	800.00	- 850.00	0	0.00	24	96.00
18	850.00	- 900.00	0	0.00	24	96.00
19	900.00	- 950.00	0	0.00	24	96.00
20	950.00	- 1000.00	0	0.00	24	96.00
21	1000.00	- 1050.00	1	4.00	25	100.00

Enter the name of the desired variable: @

Press any key to continue. (The screen will clear.)

Figure 4.2 Screen display of summary statistics and frequency distribution of silver values for data file BUTTE-2.

```
PROGRAM 4-1 SUMMARY STATISTICS

Please wait for subroutines to load.

Processing begun at 08:57:58 on 08-21-1986

Press any key to continue. (The screen will clear.)

PART 1. NAME THE DATA FILE

(A) Enter the name of the data file: BUTTE-2

Press any key to continue. (The screen will clear.)

PART 2. FILE DESCRIPTION

Header lines

Data file BUTTE-2
Constructed from file BUTTE
21 August 1986
```

Press any key to continue. (The screen will clear.)

PART 3. FILE STRUCTURE

(A) The file contains this (these) identification
    and location variable(s):

Station ID
Geology
Lithology
  X-coord.
  Y-coord.

(B) The file contains this (these) chemical element(s):

Ag
Au
Cu
Mn
Pb
Zn

(C) The file contains 265 station(s).
    Press any key to continue. (The screen will clear.)

(continued)

Figure 4.2 *(continued)*

PART 4. SELECT VARIABLES

Do you want a tabulation for Ag? (Y/N): Y
Enter the number of decimal places for output (0-3): 2

Do you want a tabulation for Au? (Y/N): N

Do you want a tabulation for Cu? (Y/N): N

Do you want a tabulation for Mn? (Y/N): N

Do you want a tabulation for Pb? (Y/N): N

Do you want a tabulation for Zn? (Y/N): N

Press any key to continue. (The screen will clear.)

PART 5. SPECIFY PARAMETERS FOR THE FREQUENCY DISTRIBUTION

Enter the percentage risk level: 10

For Ag:
a) Enter the starting value: 0
b) Enter the class interval for the frequency distribution: 10
c) Enter the detection limit: 2

Enter 'A' for an ascending, or 'D' for a descending
relative-cumulative-frequency distribution: D

Calculating. Please wait...

Press any key to continue. (The screen will clear.)

PART 6. SPECIFY THE OUTPUT DEVICE

(A) Do you want the output routed to printer? (Y/N): N

Press any key to continue. (The screen will clear.)

(continued)

Figure 4.2 (continued)

```
Data file BUTTE-2
Constructed from file BUTTE
21 August 1986
```

Vari-able	No. of Samples		Mean	Confidence Limit		Standard Deviation	Coef. of Variation
	Total	>DL*		Lower	Upper		
Ag	265	18	11.72	5.72	17.73	14.65	1.25

* > DL = Greater than detection limit

Press any key to continue. (The screen will clear.)

PART 7. DETAILED PRINTER OUTPUT

Do you want detailed output to the printer (Y/N): Y

Turn on the printer and set the paper.

Press any key when ready.

Press @ when you have entered all your data

For detailed output to the printer,
Enter the name of the desired variable: **Ag**

Data file BUTTE-2
Constructed from file BUTTE
21 August 1986

+++VARIABLE: **Ag**

Total observations = 265   Number above detection limit = 18

T(10)% = 1.74

Mean	Variance	Standard Deviation	Coefficient of Variation
11.72	214.57	14.65	1.25

Confidence Limits	Lower	Upper
Mean	5.72	17.73
Variance	132.21	420.70

Minimum value: 5.00
Maximum value: 69.00

(continued)

Figure 4.2 (continued)

FREQUENCY DISTRIBUTION

18 values from 265 observations
Class interval = 10

Number	Class Interval	Frequency	Relative Freq.(%)	Cumulative Frequency	Relative Cum. Freq. (%)
8	70.00 – 80.00	1	5.56	1	5.56
7	60.00 – 70.00	0	0.00	1	5.56
6	50.00 – 60.00	0	0.00	1	5.56
5	40.00 – 50.00	0	0.00	1	5.56
4	30.00 – 40.00	0	0.00	1	5.56
3	20.00 – 30.00	1	5.56	2	11.11
2	10.00 – 20.00	16	88.89	18	100.00
1	0.00 – 10.00	0	0.00	18	100.00

Enter the name of the desired variable: @

Press any key to continue. (The screen will clear.)

5

Making Histograms

The frequency tables you prepared in Chapter 4 provide summaries of your data files. Because it is even easier to look at a picture than to read a table, you generally will want to draw a *histogram* to represent a frequency table pictorially.

5.1 OVERVIEW

Program 5-1 prepares histograms that can be displayed either on the screen or on the printer. You may need to experiment to make a histogram to display a data set clearly. The histograms are scaled to be easy to read; therefore high values will not be displayed if you select class intervals that are too large.

5.2 BEGINNING TUTORIAL

The beginning tutorial teaches you to make a histogram for the zinc observations in data file NORWAY-2. You will learn to change parameters until you get a meaningful histogram.

Program 5-1 has these seven parts:

1. Name the data file.

2. File description.

3. File structure.

4. Specify the output device.

5. Select a variable.

6. Specify parameters for the histogram.

7. Change parameters for this histogram (if you wish).

Following are the step-by-step instructions for the sample run on data file NORWAY-2 (Figure 5.1*):

1. Parts 1 to 4 follow the form of previous programs.

2. In Part 5, you select the variable **Zn**.

3. In Part 6, you respond to queries (A) through (G) to enter parameters for the histogram. To demonstrate the steps you will go through to construct a meaningful histogram, I entered the values listed in Figure 5.1.

4. The histogram contains only 24 values because the range, as calculated from the class interval I specified in Part 6(E), was too small. (Because the range of values is provided by Program 4-1 [Figure 4.1], I could have calculated this range had I not wanted to show how to find it empirically.)

5. In Part 7, I entered **Y** in response to the query about changing parameters for the histogram. The program reverted to Part 6.

6. Rerunning the program, I entered **Y** in response to the query about plotting natural logarithms (Part 6[B]).

7. In Part 6(E), I changed the class interval to .2. Plotting logarithms allows the histogram to include all values. Alternatively, you could double the class interval and plot the original values.

5.3 EXTENDED TUTORIAL

Figure 5.2 gives the results of running Program 5-1 on data file BUTTE-2 to get output for both original observations and natural logarithms. For both plots, the one extreme value of Cu is not plotted. Scaling the plot to include this value would make it so compressed that the overall pattern of data would be obscured.

*All figures appear at the end of the chapter.

5.4 COMMENTS

In Part 6(C), the comments about decimal points in Section 3.5 also apply. In Part 6(D), you can specify a negative starting value if you wish. If you specify unreasonable values for the plot, none is displayed, and an error message is printed.

Figure 5.1   Screen display for making a histogram of zinc observations in data file NORWAY-2.

```
PROGRAM 5-1 TO MAKE A HISTOGRAM

Please wait for subroutines to load.

Processing begun at 09:13:50 on 08-21-1986

Press any key to continue. (The screen will clear.)

PART 1. NAME THE DATA FILE

(A) Enter the name of the data file: NORWAY-2

Press any key to continue. (The screen will clear.)

PART 2. FILE DESCRIPTION

Header lines

File NORWAY-2
Constructed from Table 1.1
20 August 1986

Press any key to continue. (The screen will clear.)
```

PART 3. FILE STRUCTURE

(A) The file contains this (these) identification
    and location variable(s):

Station ID

(B) The file contains this (these) chemical element(s):

Zn
Fe
Mn
Cd
Cu
Pb

(C) The file contains 25 station(s).
    Press any key to continue. (The screen will clear.)

PART 4. SPECIFY THE OUTPUT DEVICE

(A) Do you want the output routed to printer? (Y/N): N

    Press any key to continue. (The screen will clear.)

(continued)

Figure 5.1 (*continued*)

PART 5. SELECT A VARIABLE

(A) For which variable do you want to plot a histogram? Zn

Press any key to continue. (The screen will clear.)

PART 6. SPECIFY PARAMETERS FOR THE HISTOGRAM

Parameters for: Zn

(A) Enter a one-line title (70 or fewer characters) or press the
    carriage-return key if you do not want a title:

Zinc data from file NORWAY-2

(B) Take natural logarithms of the observations? (Y/N): N

(C) Enter number of decimal places for Zn (0-4): 2

(D) Enter the starting value: 0

(E) Enter the class interval for the histogram: 10

(F) Enter the detection limit: 0

(G) Ignore values below the detection limit? (Y/N): Y

Calculating. Please wait...

Press any key to continue. (The screen will clear.)

Zinc data from file NORWAY-2

24 observations within the range from 25 total.

Histogram for Zn (Original observations)

```
Count
   ----    *
  5+       * *
   ----    * *
          ***
          ***  **  *
          **** **  **                          *                         * *  --
   0+ ----+--------+--------+--------+--------+--------+--------+
   0.00    100.00   200.00   300.00   400.00   500.00
```

Press any key to continue. (The screen will clear.)

(continued)

Figure 5.1 (continued)

PART 7. CHANGE PARAMETERS FOR THIS HISTOGRAM

(A) Do you want to change any parameter for this histogram? (Y/N): Y

PART 6. SPECIFY PARAMETERS FOR THE HISTOGRAM

Parameters for: Zn

(A) Enter a one-line title (70 or fewer characters) or press the
    carriage-return key if you do not want a title:

Zinc data from file NORWAY-2

(B) Take natural logarithms of the observations? (Y/N): Y
Any negative or zero observations will be ignored.

(C) Enter number of decimal places for Zn (0-4): 2

(D) Enter the starting value: 0

(E) Enter the class interval for the histogram: .2

(F) Enter the detection limit: 0

(G) Ignore values below the detection limit? (Y/N): Y

Calculating. Please wait...

Press any key to continue. (The screen will clear.)

Zinc data from file NORWAY-2

25 observations within the range from 25 total.

Histogram for Zn (Natural logarithms)

```
Count
5+
       |                    *
       |                    *
       |                  ** ** *
       |                  *****  ** *
       |                  *****  ***          *    *
       |                                      *    *      *
      0+----+-----+-----+-----+-----+-----+-----+-----+----
     0.00  2.00      4.00      6.00
```

Press any key to continue. (The screen will clear.)

PART 7. CHANGE PARAMETERS FOR THIS HISTOGRAM

(A) Do you want to change any parameter for this histogram? (Y/N) : N

Press any key to continue. (The screen will clear.)

Figure 5.2 Screen display for making a histogram of original observations and natural logarithms of copper observations in data file BUTTE-2.

```
PROGRAM 5-1 TO MAKE A HISTOGRAM

Processing begun at 09:20:54 on 08-21-1986

Press any key to continue. (The screen will clear.)

PART 1. NAME THE DATA FILE

(A) Enter the name of the data file: BUTTE-2

Press any key to continue. (The screen will clear.)

PART 2. FILE DESCRIPTION

Header lines

Data file BUTTE-2
Constructed from file BUTTE
21 August 1986

Press any key to continue. (The screen will clear.)
```

PART 3. FILE STRUCTURE

(A) The file contains this (these) identification
    and location variable(s):

Station ID
Geology
Lithology
  X-coord.
  Y-coord.

(B) The file contains this (these) chemical element(s):

Ag
Au
Cu
Mn
Pb
Zn

(C) The file contains 265 station(s).
    Press any key to continue. (The screen will clear.)

(continued)

Figure 5.2 *(continued)*

PART 4. SPECIFY THE OUTPUT DEVICE

(A) Do you want the output routed to printer? (Y/N): N

Press any key to continue. (The screen will clear.)

PART 5. SELECT A VARIABLE

(A) For which variable do you want to plot a histogram? Cu

Press any key to continue. (The screen will clear.)

PART 6. SPECIFY PARAMETERS FOR THE HISTOGRAM

Parameters for: Cu

(A) Enter a one-line title (70 or fewer characters) or press the
    carriage-return key if you do not want a title:

(B) Take natural logarithms of the observations? (Y/N): N

(C) Enter number of decimal places for Cu (0-4): 0

(D) Enter the starting value: 0

(E) Enter the class interval for the histogram: 100

(F) Enter the detection limit: 0

(G) Ignore values below the detection limit? (Y/N): Y

Calculating. Please wait...

Press any key to continue. (The screen will clear.)

(continued)

Figure 5.2 (continued)

264 observations within the range from 265 total.

Histogram for Cu (Original observations)

```
Count
185+**
   **
   **
   **
123+**
   **
   **
   **
62+***
   ***
   **
   **
   ******  * **  *                    *                    *
 0+--------+---------+---------+---------+---------+
   0      1000      2000      3000
```

Press any key to continue. (The screen will clear.)

PART 7. CHANGE PARAMETERS FOR THIS HISTOGRAM

(A) Do you want to change any parameter for this histogram? (Y/N): Y

PART 6. SPECIFY PARAMETERS FOR THE HISTOGRAM

Parameters for: Cu

(A) Enter a one-line title (70 or fewer characters) or press the
carriage-return key if you do not want a title:

(B) Take natural logarithms of the observations? (Y/N): Y
Any negative or zero observations will be ignored.

(C) Enter number of decimal places for Cu (0-4): 0

(D) Enter the starting value: 0

(E) Enter the class interval for the histogram: .2

(F) Enter the detection limit: 0

(G) Ignore values below the detection limit? (Y/N): Y

Calculating. Please wait...

Press any key to continue. (The screen will clear.)

(continued)

Figure 5.2 (continued)

264 observations within the range from 265 total.

Histogram for Cu (Natural logarithms)

```
Count
 39+ -------------------------------------
                       *
                       *
                       *
                      **
                     ***
 26+ ----------*-----*****----------------
            *  *   ******
            ******  ******
            ************
            *************
            **************
 13+ -------***************----------------
            ****************
            *****************
            ******************
            *********************
            ***********************
  0+---------+---------+---------+---------+--
          * *
  0       2         4         6         8
```

Press any key to continue. (The screen will clear.)

6

Working with Logarithms

Many data sets in exploration geochemistry contain variables that are distributed lognormally, meaning that the logarithms of the observations rather than the observations themselves follow a normal distribution more or less closely. Indeed, for many data sets all of the geochemical variables may be distributed lognormally.

Therefore, you will need to work with logarithms. Program 6-1 does a lognormal analysis, and several other programs let you use logarithms rather than original observations, if you wish.

6.1 OVERVIEW

Program 6-1 performs the calculations that Link and I describe in detail elsewhere.[1] These calculations emphasize the estimation of population means, for instance, the average grade of a gold ore body. In exploration geochemistry, unbiased estimation of means is seldom as essential as in mining geology; often we need only find trends in geochemical variables rather than absolute values. Nonetheless, we can surmise from the coefficients of variation calculated in this program how skewed the fitted lognormal distributions are, and if they are indeed appropriate.

In this program, Part 5(C) allows you to select a positive constant to add to each observation. The purpose of doing this is to transform a nearly lognormal distribution to a normal one; the method was proposed by Krige.[2]

6.2 BEGINNING TUTORIAL

You can learn to use program 6-1 by practicing on file WELKOM, which is a set of 13 observations from the Welkom mine in South Africa.[4] Your results, except for round-off errors, will match those in the *Statistical Analysis of Geological Data*.[5]

Program 6-1 has six parts:

1. Name the data file.

2. File description.

3. File structure.

4. Specify the output device.

5. Select a variable.

6. Repeat the analysis (if you wish).

Following are the step-by-step instructions for processing the Welkom data file (Table 6.1) for the sample run (Figure 6.1*).

1. Parts 1 through 4 are the same as for previous programs.

2. In Part 5, you select a variable, which is **Au** for the Welkom example. Then you specify the number of decimal places; three places give enough detail to verify

TABLE 6.1
Gold Assays from the Welkom Mine, South Africa

Au^a
154
525
1560
1252
377
70
308
109
1221
15
48
237
68

Source: From G. S. Koch, Jr. and R. F. Link, 1980, *Statistical Analysis of Geological Data*, Vol. 1, Dover Publications, New York, p. 219.
[a]Values in inch-pennyweights.

*All figures appear at the end of the chapter.

the results. Finally, you enter 0, indicating that you do not want to add a constant for this case.

3. The output is produced by the program without your intervention.

4. Part 6 of the program allows you to repeat the analysis, usually for another variable, but perhaps to change the decimal point or to add a different constant.

6.3 EXTENDED TUTORIAL

Figures 6.2 and 6.3 show the results (points for the confidence region omitted) of running Program 6-1 on file NORWAY-2 for the zinc and iron data, respectively. The coefficients of variation show that the distribution for zinc is moderately skewed and that that for iron is slightly skewed. For zinc, the mean calculated with logarithms is rather different from the arithmetic mean, so if bias is not a serious problem the logarithic mean would be preferred.

REFERENCES

1. G. S. Koch, Jr. and R. F. Link, 1980, *Statistical Analysis of Geological Data*, Vol. 1, Dover Publications, New York, pp. 213-231. (Reprinted, with corrections, from the original two volumes published in 1971-1972 by Wiley, New York.)
2. D. G. Krige, 1960, On the Departure of Ore Value Distributions from the Lognormal Model in South African Gold Mines, *Jour. South African Inst. Mining Metall.* 61:231-244.
3. G. S. Koch, Jr. and R. F. Link, 1980, *Statistical Analysis of Geological Data*, Vol. 2, Dover Publications, New York, pp. 381-389. (Reprinted, with corrections, from the original two volumes published in 1971-1972 by Wiley, New York.)
4. Ibid., p. 219.
5. Ibid., pp. 219-228.

Figure 6.1   Screen display for a logarithmic analysis of data file WELKOM.

```
PROGRAM 6-1 WORKING WITH LOGARITHMS

Processing begun at 10:05:17 on 08-21-1986

  Press any key to continue. (The screen will clear.)

PART 1. NAME THE DATA FILE

(A) Enter the name of the data file: WELKOM

  Press any key to continue. (The screen will clear.)
```

PART 2. FILE DESCRIPTION

Header lines

Welkom mine gold assays
From Koch and Link, 1970, Statistical Analysis, v.1, p. 219
21 August 1986

Press any key to continue. (The screen will clear.)

PART 3. FILE STRUCTURE

(A) The file contains this (these) identification
    and location variable(s):

No identification variables are present.

(B) The file contains this (these) chemical element(s):

Au

(C) The file contains   13   station(s).
    Press any key to continue. (The screen will clear.)

(continued)

Figure 6.1 *(continued)*

PART 4. SPECIFY THE OUTPUT DEVICE

(A) Do you want the output routed to printer? (Y/N): N

Press any key to continue. (The screen will clear.)

PART 5. SELECT A VARIABLE

(A) Which variable do you want to analyze? Au

(B) Enter number of decimal places for Au (0-4): 3

(C) Enter constant to add to each observation (may be 0): 0

WARNING: Zero or negative values
(after adding the constant) will be ignored.

Press any key to continue. (The screen will clear.)

Calculating. Please wait...

Analysis for Au

	Original Observations	Loga- rithms
Mean	457.231	5.366
Variance	282550.200	1.983
Standard deviation	531.555	1.408
Geometric mean	214.066	
Multiplying factor for the geometric mean	4.565	
Estimate of the mean from logarithms	507.383	
Estimate of the variance from logarithms	748363.700	
Coefficient of variation, C(1)	2.503	
C(2)	1.163	
C(3)	1.705	
Number of observations	13	

Press any key to continue. (The screen will clear.)

(continued)

Figure 6.1 *(continued)*

Points for the Confidence Region

No.	Original observations		Logarithms	
	Mean	Standard Deviation	Mean	Standard Deviation
1	603.893	4136.205	4.469	1.967
2	530.008	3045.874	4.509	1.878
3	466.105	2243.354	4.552	1.785
4	410.875	1651.769	4.597	1.686
5	363.205	1214.849	4.645	1.581
6	322.159	891.367	4.696	1.469
7	286.967	651.042	4.751	1.348
8	257.036	471.538	4.812	1.214
9	231.982	336.196	4.881	1.064
10	689.289	5619.478	4.430	2.052

Press any key to continue. (The screen will clear.)

11	787.981	7639.601	4.393	2.134
12	41490.770	284180.100	6.264	1.967
13	28293.280	162597.200	6.223	1.878
14	19254.680	92672.520	6.181	1.785
15	13072.690	52553.900	6.136	1.686
16	8850.689	29603.840	6.088	1.581
17	5971.910	16523.430	6.037	1.469
18	4012.404	9102.924	5.981	1.348
19	2681.007	4918.369	5.920	1.214
20	1777.837	2576.502	5.852	1.064
21	1163.242	11277.810	4.782	2.134
22	1717.213	16648.650	5.172	2.134
23	2535.003	24577.260	5.561	2.134
24	3742.247	36281.680	5.951	2.134
25	281.690	408.235	5.075	1.064
26	342.050	495.711	5.269	1.064
27	415.344	601.931	5.463	1.064
28	504.343	730.911	5.658	1.064
29	60737.380	495166.000	6.303	2.052
30	88774.320	860681.100	6.340	2.134

Press any key to continue. (The screen will clear.)

PART 6. REPEAT THE ANALYSIS?

Do you want to repeat the analysis? (Y/N): N

Figure 6.2 Screen display for a logarithmic analysis of zinc data in file NORWAY-2.

```
Analysis for Zn
-------------------------------------------------------------

                                          Original      Loga-
                                          Observations  rithms
                                          ------------  ------

Mean                                           139.60    4.23
Variance                                     53464.25    1.09
Standard deviation                             231.22    1.04
Geometric mean                                  68.87
Multiplying factor for the geometric mean        2.62
Estimate of the mean from logarithms           114.89
Estimate of the variance from logarithms     20756.79
Coefficient of variation,  C(1)                  1.40
                           C(2)                  1.66
                           C(3)                  1.25

Number of observations                             25
```

Figure 6.3 Screen display for a logarithmic analysis of iron data in file NORWAY-2.

```
Analysis for Fe
------------------------------------------------
                                   Original    Loga-
                                   Observations rithms
                                   ------------------------------------

Mean                                  2.26       0.67
Variance                              2.09       0.26
Standard deviation                    1.44       0.51
Geometric mean                        1.96
Multiplying factor for the geometric mean   1.27
Estimate of the mean from logarithms  2.22
Estimate of the variance from logarithms   1.41
Coefficient of variation, C(1)        0.55
                          C(2)        0.64
                          C(3)        0.54

Number of observations                  25
```

CHAPTER **7**

Sorted Lists

You often will want to sort a data file on one or several variables in ascending or descending order. One purpose is to group together all values above a threshold so that you can plot them or study the associated variables. Another purpose is to identify erroneous values, for instance, those that are higher than the expected or possible range for a particular element.

7.1 OVERVIEW

Program 7-1 sorts a data file on any variable you select in either ascending or descending order. The other variables are rearranged in the new file in the order of the sorted variable. For instance, when file NORWAY-2 is sorted in descending order in Section 7.2 to produce file NORWAY-3 (Table 7.1), sample number 15 and all the elemental analyses for this sample appear at the top of the list because sample 15 has the largest value of zinc, 1010 ppm.

7.2 BEGINNING TUTORIAL

For practice, sort data file NORWAY-2 (Table 1.1); Figure 7.1 is the record of a sample run sorting zinc in ascending order.*

*All figures appear at the end of the chapter.

111

TABLE 7.1
Data File NORWAY-3 Sorted on Zn

Station ID	Zn ppm	Fe pct	Mn ppm	Cd ppm	Cu ppm	Pb ppm
8	23	0.93	260	0.4	7	7
1	24	1.08	330	0.4	7	5
2	25	1.18	420	0.3	9	7
7	26	1.08	530	0.5	10	7
6	29	1.06	510	0.6	10	8
11	31	1.51	350	0.4	11	4
20	33	1.94	490	1.0	20	12
19	36	1.28	410	0.2	7	10
3	42	2.06	910	0.6	12	6
18	44	1.70	710	0.4	11	16
21	45	1.79	260	0.8	27	12
17	48	1.43	590	0.4	8	8
14	48	1.71	570	0.3	17	9
4	50	1.73	700	0.5	15	9
5	52	1.74	690	0.5	18	10
10	72	3.35	530	0.5	11	8
25	80	3.61	900	1.0	23	10
24	81	2.62	970	0.9	22	13
9	89	1.84	670	0.9	32	6
12	115	1.59	650	1.2	37	6
22	118	6.70	1930	1.2	33	17
23	274	6.10	5920	3.1	63	27
13	535	2.59	960	3.5	350	9
16	560	3.05	450	3.1	490	8
15	1010	2.71	1070	5.6	590	9

Program 7.1 has 5 parts:

1. Name the data file.

2. File description.

3. File structure.

4. Name the sorted data file.

5. Select the variable to sort.

Following are the step-by-step instructions:
1. Parts 1 through 3 follow the same form as for previous programs.
2. In Part 4, you name the sorted data file; I entered **NORWAY-3**. Then the

TABLE 7.2
Part of File BUTTE-3 Sorted on Ag

Station ID	Ag ppm	Au ppm	Cu ppm	Mn ppm	Pb ppm	Zn ppm
M24399	69	2	132	1236	2	131
M24101	16	2	31	1546	2	130
M24196	14	2	68	1517	12	101
M24481	13	2	22	1385	2	68
M25311	10	2	396	5321	1818	1969
M24120	9	2	44	556	2	99
M24441	9	2	94	1074	14	25
M24126	9	2	45	935	2	25
M24443	8	2	168	1343	9	25
M24161	7	440	71	1123	81	86
M24099	7	2	35	1294	2	25
M24321	7	50	13	1234	2	66
M24330	7	2	136	1107	8	67
M25345	6	2	3617	624	518	996
M24125	5	2	97	1021	12	25
M24315	5	2	46	1423	12	137
M24312	5	47	42	1173	2	36
M24100	5	2	35	1296	2	25

program issues a warning indicating the amount of disk space that you need to save the new file. This amount is the disk space of the existing file, which is 3712 bytes for file NORWAY-2. Unless your diskette has this much space, the sorted file cannot be written to it. One way to avoid this problem is always to write the sorted file to a second diskette if you have two disk drives. Because there was enough space on the diskette, I entered **Y** to continue. Finally, I entered carriage-returns to keep the original titles.

3. In Part 5(A), I entered the name of the variable to sort as **Zn**, and entered **A** to indicate that the sort was to be in ascending order.

To list the sorted file and produce Table 7.1, I used Program 3.1.

7.3 EXTENDED TUTORIAL

One reason for sorting data files is to group together the anomalous or possibly anomalous observations. Table 7.2 is a partial listing of a data file made using Program 7-1 to sort data file BUTTE-2 on silver values in descending order. The table lists all silver values above the detection limit; some of the stations with high silver values also have high values for the other elements.

Figure 7.1 Screen display of sorting data file NORWAY-2 in ascending order.

```
PROGRAM 7-1 TO SORT A DATA FILE

Please wait for subroutines to load.

Processing begun at 10:57:51 on 08-21-1986

Press any key to continue. (The screen will clear.)

PART 1. NAME THE DATA FILE

(A) Enter the name of the data file: NORWAY-2

Press any key to continue. (The screen will clear.)

PART 2. FILE DESCRIPTION

Header lines

File NORWAY-2
Constructed from Table 1.1
20 August 1986
```

Press any key to continue. (The screen will clear.)

PART 3. FILE STRUCTURE

(A) The file contains this (these) identification
    and location variable(s):

Station ID

(B) The file contains this (these) chemical element(s):

Zn
Fe
Mn
Cd
Cu
Pb

(C) The file contains 25 station(s).
    Press any key to continue. (The screen will clear.)

(continued)

Figure 7.1 *(continued)*

PART 4. NAME THE SORTED DATA FILE

(A) Enter the name for the sorted data file: NORWAY-3

WARNING: You must have at least 3712 bytes free
         on your disk

(C) Do you want to continue? (Y/N): Y

(D) Enter three lines of title information
    or press the carriage-return to keep the old titles

Enter line 1 : File NORWAY-3
Enter line 2 : From file NORWAY-2 sorted on Zn in ascending order
Enter line 3 : 21 August 1986
Press any key to continue. (The screen will clear.)

PART 5. SELECT THE VARIABLE TO SORT

(A) Which variable do you want to sort? Zn

(B) Enter 'A' for an ascending or 'D' for a descending
    sorting: A

Reading data. Please wait...

Press any key to continue. (The screen will clear.)

Sorting. Please wait...

Pass 1 of 4 Pass 2 of 4 Pass 3 of 4 Pass 4 of 4

Writing the sorted data file. Please wait...

CHAPTER 8

Estimating
Geochemical
Thresholds

Estimating geochemical thresholds is an essential step in most analyses of exploration-geochemical data. The standard textbooks explain the various methods that have been used. For procedures, read Rose, Hawkes, and Webb and Levinson.[1] Working with Program 4.1, you can use the method of selecting a threshold as a percentage point of the frequency distribution.

This section presents a method devised by Miesch, who writes that

A statistic is proposed for estimating the geochemical threshold and its statistical significance, or it may be used to identify a group of extreme values that can be tested for significance by other means. The statistic is the maximum gap between adjacent values in an ordered array after each gap has been adjusted for the expected frequency. The values in the ordered array are geochemical values transformed by either $ln(x - a)$ or $ln(a - x)$ and then standardized so that the mean is zero and the variance is unity. The expected frequency is taken from a fitted normal curve with unit area. The midpoint of an adjusted gap that exceeds the corresponding critical value may be taken as an estimate of the geochemical threshold, and the associated probability indicates the likelihood that the threshold separates two geochemical populations.[2]

8.1 OVERVIEW

Program 8-1 allows you to search either the entire range of values or just the values above the median, which would usually be the preference.

TABLE 8.1

Uranium Concentrations in Samples of Ground Water from the Carrizo Area of Southern Texas

0.003	Downdip regime	0.04	Transition zone
0.003	Transition zone	0.05	Downdip regime
0.003	Downdip regime	0.12	Transition zone
0.004	Downdip regime	0.20	Outcrop regime
0.004	Downdip regime	0.20	Transition zone
0.004	Downdip regime	0.21	Outcrop regime
0.004	Downdip regime	0.25	Transition zone
0.005	Downdip regime	0.30	Outcrop regime
0.006	Downdip regime	0.32	Outcrop regime
0.006	Transition zone	0.44	Outcrop regime
0.006	Transition zone	0.51	Outcrop regime
0.007	Downdip regime	0.51	Outcrop regime
0.008	Downdip regime	0.59	Outcrop regime
0.008	Downdip regime	0.63	Outcrop regime
0.009	Transition zone	0.80	Outcrop regime
0.01	Downdip regime	1.14	Outcrop regime
0.01	Transition zone	1.20	Outcrop regime
0.02	Transition zone	1.49	Outcrop regime
0.02	Downdip regime	1.64	Outcrop regime
0.02	Transition zone	1.70	Outcrop regime
0.03	Downdip zone	2.56	Outcrop regime
0.03	Transition zone	3.30	Outcrop regime
0.03	Transition zone	3.53	Outcrop regime
0.04	Transition zone		

Source: From A. T. Miesch, 1981, Estimation of the Geochemical Threshold and Its Statistical Significance, *Jour. Geochem. Exploration* **16**:63.
Note: Values in micrograms per liter.

8.2 BEGINNING TUTORIAL

You can learn to use the program by applying it to Miesch's test data,[3] which are in the file CARRIZO. The data—47 observations of uranium concentrations in ground water from the Carrizo area of southern Texas—are listed in Table 8.1.

Program 8-1 has six parts:

1. Name the data file.

2. File description.

3. File structure.

4. Specify the output device.

5. Select a variable.

6. Display results of the analysis on the printer or screen.

120

Following are the step-by-step instructions for the sample run illustrated in Figure 8.1:*
1. Parts 1 through 4 follow the same form as for previous programs.
2. In Part 5, select the variable **U** for analysis. (You must precede the character **U** by a blank character, obtained by depressing the space bar, because the variable name has only one character.) In order to get Miesch's output,[4] type **Y** in response to the 5(B) query to select the entire range of observations.
3. The output matches Miesch's.[5]

8.3 EXTENDED TUTORIAL

Applying Program 8-1 to data file NORWAY-2 (Table 1.1), and selecting zinc observations above the median for analysis, provides the results in Figure 8.2. Note that the gap statistic is smaller than the critical value, indicating that it is not significant at the 10-percent risk level.

REFERENCES

1. A. W. Rose, H. E. Hawkes, and J. S. Webb, 1979, *Geochemistry in Mineral Exploration* (2nd ed.), Academic Press, London, p. 34; A. A. Levinson, 1974, *Introduction to Exploration Geochemistry*, Applied Publishing Ltd., Calgary, p. 214.
2. A. T. Miesch, 1981, Estimation of the Geochemical Threshold and Its Statistical Significance, *Jour. Geochem. Exploration*, **16**:49.
3. Ibid., p. 63.
4. Ibid., p. 75.
5. Ibid., p. 75.

*All figures appear at the end of the chapter.

Figure 8.1 Screen display for estimating geochemical thresholds in data file CARRIZO.

```
PROGRAM 8-1 ESTIMATING GEOCHEMICAL THRESHOLDS

Processing begun at 08:47:15 on 08-22-1986

Press any key to continue. (The screen will clear.)

PART 1. NAME THE DATA FILE

(A) Enter the name of the data file: CARRIZO

Press any key to continue. (The screen will clear.)

PART 2. FILE DESCRIPTION

Header lines

Data file CARRIZO
from Miesch, 1981, p. 63
21 August 1986

Press any key to continue. (The screen will clear.)
```

PART 3. FILE STRUCTURE

(A) The file contains this (these) identification
    and location variable(s):

No identification variables are present.

(B) The file contains this (these) chemical element(s):

U

(C) The file contains 47 station(s).
    Press any key to continue. (The screen will clear.)

PART 4. SPECIFY THE OUTPUT DEVICE

(A) Do you want the output routed to printer? (Y/N): N

    Press any key to continue. (The screen will clear.)

PART 5. SELECT VARIABLES

(A) Which variable do you want to analyze? U

(continued)

Figure 8.1 *(continued)*

NOTE: You can either search the entire range or above the median

(B) Search the entire range? (Y/N): Y

WARNING: Values less than or equal to zero will be ignored

(C) Enter the percent risk level (0.1% to 20%): 10

Reading data. Please wait...

  Press any key to continue. (The screen will clear.)

Calculating. Please wait...

Sorting...

Pass 1 of 5 Pass 2 of 5 Pass 3 of 5 Pass 4 of 5 Pass 5 of 5

PART 6. RESULTS OF THE ANALYSIS

ANALYSIS FOR U

There are 47 samples. Searched the entire range

The data were transformed by LN(X-ALPHA) with ALPHA= 0
Skewness before transformation= 2.404517 After= .2257189

The largest adjusted gap is between the values of
     0.0500 and   0.1200

There are 21 stations above the gap

The gap statistic is   0.1521

The critical value (10 %) is 0.12110

     Press any key to continue. (The screen will clear.)

Figure 8.2   Screen display for estimating geochemical thesholds in data file NORWAY-2.

```
PROGRAM 8-1 ESTIMATING GEOCHEMICAL THRESHOLDS

Processing begun at 08:50:56 on 08-22-1986

Press any key to continue. (The screen will clear.)

PART 1. NAME THE DATA FILE

(A) Enter the name of the data file: NORWAY-2

Press any key to continue. (The screen will clear.)

PART 2. FILE DESCRIPTION

Header lines

File NORWAY-2
Constructed from Table 1.1
20 August 1986

Press any key to continue. (The screen will clear.)
```

PART 3. FILE STRUCTURE

(A) The file contains this (these) identification
    and location variable(s):

Station ID

(B) The file contains this (these) chemical element(s):

Zn
Fe
Mn
Cd
Cu
Pb

(C) The file contains 25 station(s).
    Press any key to continue. (The screen will clear.)

PART 4. SPECIFY THE OUTPUT DEVICE

(A) Do you want the output routed to printer? (Y/N): N

    Press any key to continue. (The screen will clear.)

(continued)

Figure 8.2 (continued)

PART 5. SELECT VARIABLES

(A) Which variable do you want to analyze? Cu

NOTE: You can either search the entire range or above the median

(B) Search the entire range? (Y/N): N

WARNING: Values less than or equal to zero will be ignored

(C) Enter the percent risk level (0.1% to 20%): 10

Reading data. Please wait...

   Press any key to continue. (The screen will clear.)

Calculating. Please wait...

Sorting...

Pass 1 of 4 Pass 2 of 4 Pass 3 of 4 Pass 4 of 4

PART 6. RESULTS OF THE ANALYSIS

ANALYSIS FOR Cu

There are 25 samples. Searched above median on LY
The data were transformed by LN(X-ALPHA) with ALPHA= 6.624
Skewness before transformation= 2.687804 After= .3603473

The largest adjusted gap is between the values of
     63.0000 and   349.9999

There are 3 stations above the gap

The gap statistic is   0.1464

The critical value (10 %) is 0.16400

     Press any key to continue. (The screen will clear.)

CHAPTER 9

Comparing Paired
Observations

We often want to compare a pair of variables to learn if they plot on an approximately straight line or, in other words, if they are related through a simple linear regression. The method is explained in *Statistical Analysis of Geological Data;*[1] you will also find an explanation in standard statistical textbooks.

9.1 OVERVIEW

Selecting Program 9-1 from the menu (Figure 1.2), you can process a data file to compare one pair of geochemical variables, several pairs, or all pairs, depending on your purpose and on how many variables are present in the file.

9.2 BEGINNING TUTORIAL

File FRESNI contains the sample set of data, which are silver assays taken at 2-meter intervals in a vein mined at Fresnillo, Mexico. The data are listed in Table 9.1.
Program 9-1 has seven parts:

1. Name the data file.

2. File description.

3. File structure.

4. Specify the output device.

TABLE 9.1
Observations of Silver Content at 22 Contiguous
Sample Points in a Drift in the Fresnillo Mine,
Zacatecas, Mexico

X-coord.	Ag[a]
2	698
4	365
6	223
8	335
10	156
12	512
14	357
16	274
18	454
20	369
22	179
24	194
26	137
28	40
30	65
32	16
34	100
36	22
38	13
40	19
42	72
44	23

Source: From G. S Koch, Jr. and R. F. Link, 1980, *Statistical Analysis of Geological Data*, Vol. 2, Dover Publications, New York, p. 9.
[a]Units are meter-grams per metric ton.

5. Select the independent variable.

6. Select the dependent variable.

7. Repeat for another dependent variable, if desired.

Following are the step-by-step instructions for the sample run illustrated in Figure 9.1:*

 1. Parts 1 through 4 follow the same form as for previous programs.

 2. In Part 5, select the independent variable. For the silver data example, the independent variable clearly is the *x*-value of distance. For two geochemical variables, the choice of independent variable may not be obvious, as explained elsewhere.[2] You also need to specify the number of decimal places to be printed in the resulting tables and the percent risk level.

*All figures appear at the end of the chapter.

3. In Part 6, specify the dependent variable, the number of decimal places, and the percent risk level.

4. The resulting analysis follows the form of *Statistical Analysis of Geological Data;* except for roundoff errors, the results are the same.[3]

5. Finally, because the data file contains only two variables, there is no opportunity to repeat the analysis for a second variable.

9.3 EXTENDED TUTORIAL

For an extended tutorial, you can use data file NORWAY-2 (Table 1.1) with zinc as the independent variable and copper and lead as the dependent variables. Figures 9.2 and 9.3 provide scatter diagrams for the two cases, prepared using Program 10-1. Figure 9.4 shows the results of a sample run following the procedure used for the beginning tutorial. As you might have guessed from the scatter diagram, there is a strong linear relationship between zinc and copper (correlation coefficient of 0.969) but none between zinc and lead (correlation coefficient of 0.096).

REFERENCE

1. G. S. Koch, Jr. and R. F. Link 1980, *Statistical Analysis of Geological Data,* Vol. 2, Dover Publications, New York, pp. 7-18. (Reprinted, with corrections, from the original two volumes published in 1971-1972 by Wiley, New York.)
2. Ibid., p. 18.
3. Ibid., pp. 12, 13, 17.

Figure 9.1 Screen display of simple linear regression analysis of silver assays in data file FRESNI.

```
PROGRAM 9-1 TO COMPARE PAIRED OBSERVATIONS

Processing begun at 14:38:49 on 08-26-1986

Press any key to continue. (The screen will clear.)

PART 1. NAME THE DATA FILE

(A) Enter the name of the data file: FRESNI

Press any key to continue. (The screen will clear.)
```

PART 2. FILE DESCRIPTION

Header lines

Data file FRESNI
From Koch and Link, Statistical Analysis, 1971, p. 9
22 August 1986

Press any key to continue. (The screen will clear.)

PART 3. FILE STRUCTURE

(A) The file contains this (these) identification
    and location variable(s):

    X-coord.

(continued)

Figure 9.1 *(continued)*

(B) The file contains this (these) chemical element(s):

Ag

(C) The file contains 22 station(s).
Press any key to continue. (The screen will clear.)

PART 4. SPECIFY THE OUTPUT DEVICE

(A) Do you want the output routed to printer? (Y/N): N

Press any key to continue. (The screen will clear.)

PART 5. SELECT THE INDEPENDENT VARIABLE

In the following question,
if you want to use the X-coordinate enter 1,
otherwise enter the element name

(A) Enter the name of the independent variable: 1

(B) Enter number of decimal places for X-coord. (0-4): 0

(C) Enter the percent risk level: 10

Press any key to continue. (The screen will clear.)

PART 6. SELECT THE DEPENDENT VARIABLE

In the following question,
if you want to use the X-coordinate enter 1,
otherwise enter the element name

(A) Enter the name of the dependent variable: **Ag**

(B) Enter number of decimal places for **Ag** (0-4): 2

Press any key to continue. (The screen will clear.)

(continued)

Figure 9.1 *(continued)*

```
Calculating. Please wait...

Press any key to continue. (The screen will clear.)

|-------------------------------------------------------------|
|-------------------------------------------------------------|

                      Linear regression

|-------------------------------------------------------------|
|-------------------------------------------------------------|

The independent variable is  X-coord., with a mean value of    23
The dependent variable is Ag, with a mean value of   210.14.
The number of observations is  22 .

|-------------------------------------------------------------|

Slope (Regression coefficient) =        -11.744
Regression equation: Ag(hat) = a + b *  X-coord.
Ag(hat) =     480.247 +    -11.744 *  X-coord.
Correlation coefficient, r(Ag,  X-coord.) = -.805

|-------------------------------------------------------------|
```

Press any key to continue. (The screen will clear.)

Analysis of variance

Source of variation	Sum of squares	d.f.	Mean square	F	F(10.0%)
Regression	488512	1	488512	36.876	2.975
Residual	264946	20	13247		
Total	753459	21			

Press any key to continue. (The screen will clear.)

PART 7. REPEAT FOR ANOTHER DEPENDENT VARIABLE

(A) Do you want to perform a regression for another variable? (Y/N): N

Press any key to continue. (The screen will clear.)

(continued)

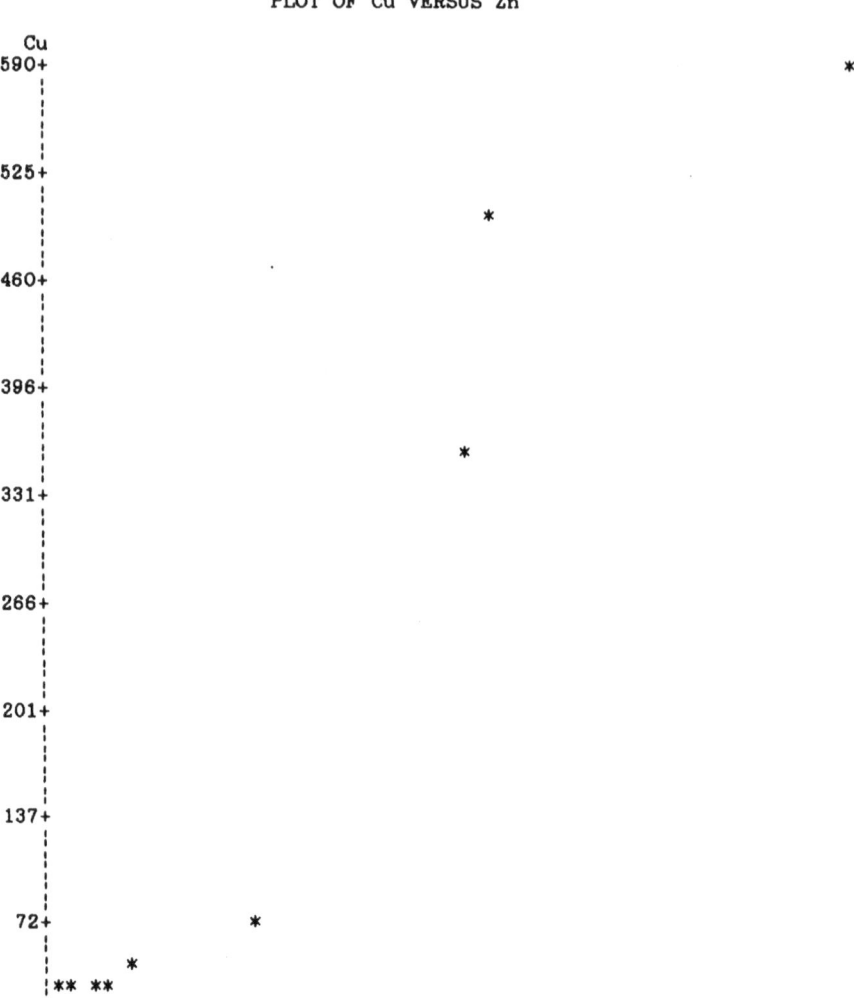

Figure 9.2 Plot of Cu vs Zn in data file NORWAY-2.

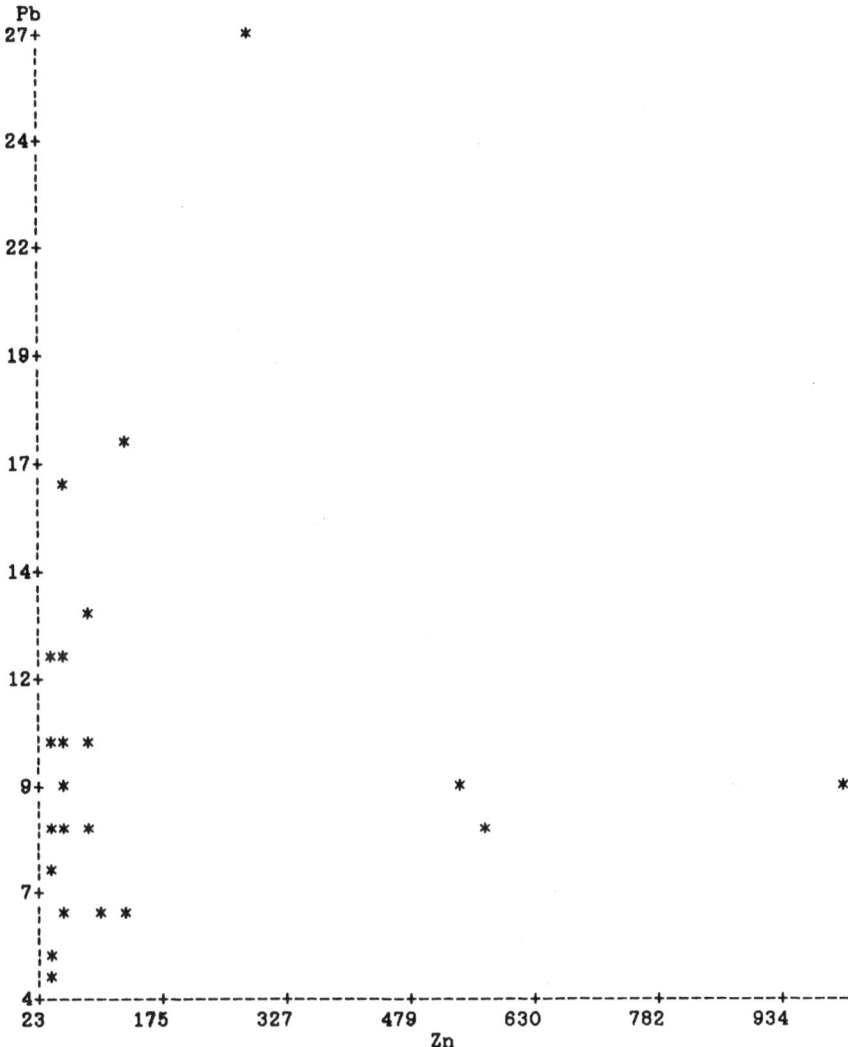

Figure 9.3 Plot of Pb vs Zn in data file NORWAY-2.

Figure 9.4   Screen display of simple linear regression analysis of zinc with copper and lead in data file NORWAY-2.

```
PROGRAM 9-1 TO COMPARE PAIRED OBSERVATIONS'

Processing begun at 09:48:36 on 08-26-1986

Press any key to continue. (The screen will clear.)

PART 1. NAME THE DATA FILE

(A) Enter the name of the data file: NORWAY-2

Press any key to continue. (The screen will clear.)

PART 2. FILE DESCRIPTION

Header lines

File NORWAY-2
Constructed from Table 1.1
20 August 1986

Press any key to continue. (The screen will clear.)
```

PART 3. FILE STRUCTURE

(A) The file contains this (these) identification
    and location variable(s):

Station ID

(B) The file contains this (these) chemical element(s):

Zn
Fe
Mn
Cd
Cu
Pb

(C) The file contains 25   station(s).
    Press any key to continue. (The screen will clear.)

PART 4. SPECIFY THE OUTPUT DEVICE

(A) Do you want the output routed to printer? (Y/N): N

    Press any key to continue. (The screen will clear.)

(continued)

Figure 9.4 *(continued)*

PART 5. SELECT THE INDEPENDENT VARIABLE

(A) Enter the name of the independent variable: Zn

(B) Enter number of decimal places for Zn (0-4): 2

(C) Enter the percent risk level: 10

Press any key to continue. (The screen will clear.)

PART 6. SELECT THE DEPENDENT VARIABLE

(A) Enter the name of the dependent variable: Cu

(B) Enter number of decimal places for Cu (0-4): 2

Press any key to continue. (The screen will clear.)

Calculating. Please wait...

Press any key to continue. (The screen will clear.)

```
                        Linear regression

The independent variable is Zn, with a mean value of    139.60.
The dependent variable is Cu, with a mean value of       73.60.
The number of observations is  25 .

Slope (Regression coefficient)  =      0.655
Regression equation: Cu(hat) = a + b * Zn
Cu(hat) =   -17.876 +   0.655 * Zn
Correlation coefficient, r(Cu,Zn) = 0.969
```

Press any key to continue. (The screen will clear.)

```
                     Analysis of variance

Source of   Sum of squares  d.f.   Mean square       F      F( 10.0%)
variation

Regression     550954.60      1      550954.60     354.447    2.937
Residual        35751.38     23        1554.41
Total          586706.00     24
```

Press any key to continue. (The screen will clear.)

(continued)

Figure 9.4 *(continued)*

PART 7. REPEAT FOR ANOTHER DEPENDENT VARIABLE

(A) Do you want to perform a regression for another variable? (Y/N): Y
Press any key to continue. (The screen will clear.)

PART 6. SELECT THE DEPENDENT VARIABLE

(A) Enter the name of the dependent variable: Pb

(B) Enter number of decimal places for Pb (0-4): 2

Press any key to continue. (The screen will clear.)

Calculating. Please wait...

Press any key to continue. (The screen will clear.)

```
                        Linear regression
------------------------------------------------------------------------

The independent variable is Zn, with a mean value of      139.60.
The dependent variable is Pb, with a mean value of          9.72.
The number of observations is   25 .

------------------------------------------------------------------------

Slope (Regression coefficient) =      0.002
Regression equation: Pb(hat) = a + b * Zn
Pb(hat) =       9.442 +    0.002 * Zn
Correlation coefficient, r(Pb,Zn) = 0.096

------------------------------------------------------------------------

Press any key to continue. (The screen will clear.)
```

Figure 9.4 *(continued)*

```
                Analysis of variance

Source of   Sum of squares   d.f.   Mean square      F      F( 10.0%)
variation

Regression        5.07         1        5.07       0.216      2.937
Residual        539.97        23       23.48
Total           545.04        24

Press any key to continue. (The screen will clear.)

PART 7. REPEAT FOR ANOTHER DEPENDENT VARIABLE

(A) Do you want to perform a regression for another variable? (Y/N): N

Press any key to continue. (The screen will clear.)
```

CHAPTER **10**

Plotting Data

Plotting data is essential for recognizing patterns and relationships among variables. A plot also will help you identify unusual values that may be anomalous or that may be mistakes.

10.1 OVERVIEW

Program 10-1 plots pairs of variables. The following options are available:

1. You can define scales or choose automatic scaling.

2. You can make a three-dimensional plot by plotting the scaled value of a third variable at each data point.

3. You can plot on either the screen or the printer.

You can combine these options, perhaps to make unscaled screen plots for editing data or to make scaled printer plots for permanent records.

So that you can plot using any standard printer, Program 10-1 is written using common ASCII characters, each one of which occupies a rectangle that measures 1/10-inch horizontally and 1/6-inch vertically. Therefore, if two or more data points fall within one of these rectangles, only one symbol appears, corresponding to the point with the largest value.

The characters are ones that will appear the same on most, if not all, IBM-compatible computer screens and printers. (There is, unfortunately, no industry standard; see Appendix B for details.)

10.2 BEGINNING TUTORIAL

Program 10.1 has eight parts:

1. Name the data file.

2. File description.

3. File structure.

4. Specify the output device.

5. Select the variable for the horizontal axis.

6. Select the variable for the vertical axis.

7. Specify the z variable.

8. Repeat for another set of variables.

Following are step-by-step instructions for the sample run to plot Cu versus Zn in data file NORWAY-2 illustrated in Figure 10.1.*

   1.   Parts 1 through 4 follow the same form as for previous programs.

   2.   In Part 5, select automatic scaling so that the plot is produced without having to define axes. (The next section explains how to scale the same set of variables.) Enter **Zn** for the horizontal axis, with zero decimal places.

   3.   In Part 6, enter **Cu** as the variable for the vertical axis, again with zero decimal places.

   4.   In Part 7, enter **N** to indicate that you do not want a three-dimensional plot.

   5.   Finally, in Part 8, enter **N** to indicate that you do not want to plot another set of variables.

10.3 EXTENDED TUTORIAL A: SCALED PLOTS

The automatically scaled plots explained in section 10.2 have three desirable characteristics; they are easy to plot, they fill the entire width of the screen, and they fill the print line of a standard printer. However, because the scales are not rounded, they may be inconvenient to read, and a plot from one data set cannot be compared directly with one from another data set. Therefore, I provide you with an option to scale your plots.

Figure 10.2 illustrates how you can make a scaled plot for the data already plotted in Figure 10.1. Because you have made the automatically scaled plot, it is easy to select meaningful scales for Figure 10.2.

Parts 1 through 4 follow the same form as for previous programs, as do sections

---

*All figures at the end of the chapter.

(A) through (C) of Part 5. In Parts 5(D) through 5(F), you enter the minimum and maximum values for Zn and the step. The step is used by the program to scale the axes. Because the step is 10, the axes are scaled in multiples of 10. You handle Part 6 in the same way as Part 5.

10.4 EXTENDED TUTORIAL B: THREE-DIMENSIONAL PLOTS

In a three-dimensional plot, each plotted point indicates the scaled value of another variable. Figure 10.3 shows how you can make a plot of Fe values present at the Cu-Zn points of the previous diagram in Figure 10.2.

Parts 1 through 6 are identical with Figure 10.2. In Part 7, you specify Fe as the variable for the z-axis.

152

Figure 10.1   Screen display of automatically scaled plot of Cu and Zn in data file NORWAY-2.

```
PROGRAM 10-1 TO MAKE PLOTS

Please wait for subroutines to load.

Processing begun at 09:28:55 on 08-25-1986

Press any key to continue. (The screen will clear.)

PART 1. NAME THE DATA FILE

(A) Enter the name of the data file: NORWAY-2

Press any key to continue. (The screen will clear.)
```

PART 2. FILE DESCRIPTION

Header lines

File NORWAY-2
Constructed from Table 1.1
20 August 1986

Press any key to continue. (The screen will clear.)

PART 3. FILE STRUCTURE

(A) The file contains this (these) identification
    and location variable(s):

Station ID

(continued)

Figure 10.1 (*continued*)

(B) The file contains this (these) chemical element(s):

Zn
Fe
Mn
Cd
Cu
Pb

(C) The file contains 25 station(s).
Press any key to continue. (The screen will clear.)

PART 4. SPECIFY THE OUTPUT DEVICE

(A) Do you want the output routed to printer? (Y/N): N

Press any key to continue. (The screen will clear.)

PART 5. SELECT THE VARIABLE FOR THE HORIZONTAL AXIS

(A) Do you want automatic scaling? (Y/N): Y

(B) Enter the variable name for the horizontal axis: Zn

(C) Enter number of decimal places for Zn (0-4): 0

Press any key to continue. (The screen will clear.)

(continued)

Figure 10.1 *(continued)*

PART 6. SELECT THE VARIABLE FOR THE VERTICAL AXIS

(A) Enter the variable name for the vertical axis: Cu

(B) Enter number of decimal places for Cu (0-4): 0
Press any key to continue. (The screen will clear.)

PART 7. SPECIFY Z VARIABLE

(A) Do you want a three-dimensional plot? (Y/N): N
Press any key to continue. (The screen will clear.)

Calculating. Please wait...

```
                PLOT OF Cu VERSUS Zn

Cu
590+ ----------------------------------------------                    *


396+                                                 *

                                                          *

201+

   7+
    **  ** *
   23      175      327      479      630      782      934
                              Zn

Press any key to continue. (The screen will clear.)

PART 8. PLOT ANOTHER SET OF VARIABLES?

(A) Do you want to plot another set of variables? (Y/N): N

Press any key to continue. (The screen will clear.)
```

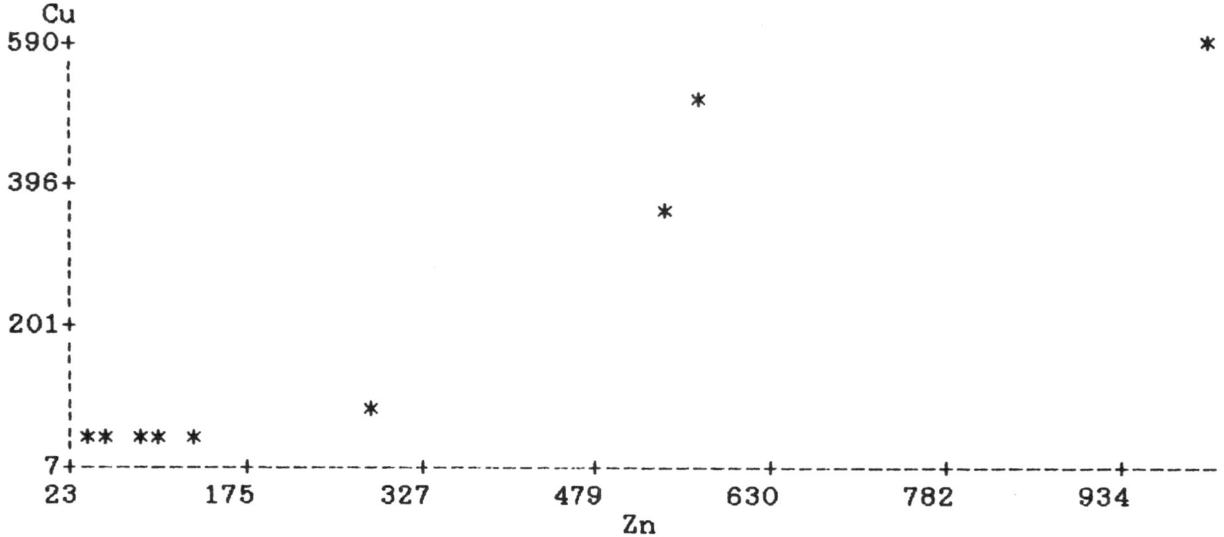

Figure 10.2 Screen display of user scaled plot of Cu and Zn in data file NORWAY-2.

```
PROGRAM 10-1 TO MAKE PLOTS

Processing begun at 09:35:03 on 08-25-1986

Press any key to continue. (The screen will clear.)

PART 1. NAME THE DATA FILE

(A) Enter the name of the data file: NORWAY-2

Press any key to continue. (The screen will clear.)
```

PART 2. FILE DESCRIPTION

Header lines

File NORWAY-2
Constructed from Table 1.1
20 August 1986

Press any key to continue. (The screen will clear.)

PART 3. FILE STRUCTURE

(A) The file contains this (these) identification
    and location variable(s):

Station ID

(continued)

Figure 10.2 *(continued)*

(B) The file contains this (these) chemical element(s):

Zn
Fe
Mn
Cd
Cu
Pb

(C) The file contains   25   station(s).
Press any key to continue. (The screen will clear.)

PART 4. SPECIFY THE OUTPUT DEVICE

(A) Do you want the output routed to printer? (Y/N): n

Press any key to continue. (The screen will clear.)

PART 5. SELECT THE VARIABLE FOR THE HORIZONTAL AXIS

(A) Do you want automatic scaling? (Y/N): N

(B) Enter the variable name for the horizontal axis: Zn

(C) Enter number of decimal places for Zn (0-4): 0

(D) Enter the minimum value for Zn: 0

(E) Enter the maximum value for Zn: 1100

(F) Enter the step for Zn: 10

Press any key to continue. (The screen will clear.)

(continued)

Figure 10.2 (continued)

PART 6. SELECT THE VARIABLE FOR THE VERTICAL AXIS

(A) Enter the variable name for the vertical axis: Cu

(B) Enter number of decimal places for Cu (0-4): 0

(C) Enter the minimum value for Cu: 0

(D) Enter the maximum value for Cu: 600

(E) Enter the step for Cu: 10

Press any key to continue. (The screen will clear.)

PART 7. SPECIFY Z VARIABLE

(A) Do you want a three-dimensional plot? (Y/N): N
Press any key to continue. (The screen will clear.)

Calculating. Please wait...

```
                PLOT OF Cu VERSUS Zn

Cu
600+
     +-------------------------------------------------
                                            *

400+
     +------

200+
     +------
                              *
                                    *
   0+                    *
     +----+-------+-------+-------+-------+-------+----
*****
     0   200     400     600     800    1000
                          Zn
```

Press any key to continue. (The screen will clear.)

PART 8. PLOT ANOTHER SET OF VARIABLES?

(A) Do you want to plot another set of variables? (Y/N): N

Press any key to continue. (The screen will clear.)

(continued)

Figure 10.3 Screen display of a three-dimensional plot of Cu, Zn, and Fe in data file NORWAY-2.

```
PROGRAM 10-1 TO MAKE PLOTS

Processing begun at 09:40:11 on 08-25-1986

   Press any key to continue. (The screen will clear.)

PART 1. NAME THE DATA FILE

(A) Enter the name of the data file: NORWAY-2

   Press any key to continue. (The screen will clear.)

PART 2. FILE DESCRIPTION

   Header lines

File NORWAY-2
Constructed from Table 1.1
20 August 1986

   Press any key to continue. (The screen will clear.)
```

PART 3. FILE STRUCTURE

(A) The file contains this (these) identification
    and location variable(s):

Station ID

(B) The file contains this (these) chemical element(s):

Zn
Fe
Mn
Cd
Cu
Pb

(C) The file contains 25 station(s).
    Press any key to continue. (The screen will clear.)

PART 4. SPECIFY THE OUTPUT DEVICE

(A) Do you want the output routed to printer? (Y/N): N

(continued)

Figure 10.3 *(continued)*

Press any key to continue. (The screen will clear.)

PART 5. SELECT THE VARIABLE FOR THE HORIZONTAL AXIS

(A) Do you want automatic scaling? (Y/N): N

(B) Enter the variable name for the horizontal axis: Zn

(C) Enter number of decimal places for Zn (0-4): 0

(D) Enter the minimum value for Zn: 0

(E) Enter the maximum value for Zn: 1100

(F) Enter the step for Zn: 10

Press any key to continue. (The screen will clear.)

PART 6. SELECT THE VARIABLE FOR THE VERTICAL AXIS

(A) Enter the variable name for the vertical axis: Cu

(B) Enter number of decimal places for Cu (0-4): 0

(C) Enter the minimum value for Cu: 0

(D) Enter the maximum value for Cu: 600

(E) Enter the step for Cu: 10

Press any key to continue. (The screen will clear.)

PART 7. SPECIFY Z VARIABLE

(A) Do you want a three-dimensional plot? (Y/N): Y

(B) Enter the variable name for the Z axis: Fe

(C) Enter number of decimal places for Fe (0-4): 2

(D) Enter the minimum value for Fe: 0

(E) Enter the maximum value for Fe: 10

(F) Enter the number of intervals (1-26) for Fe: 10

Press any key to continue. (The screen will clear.)

(continued)

Figure 10.3 (continued)

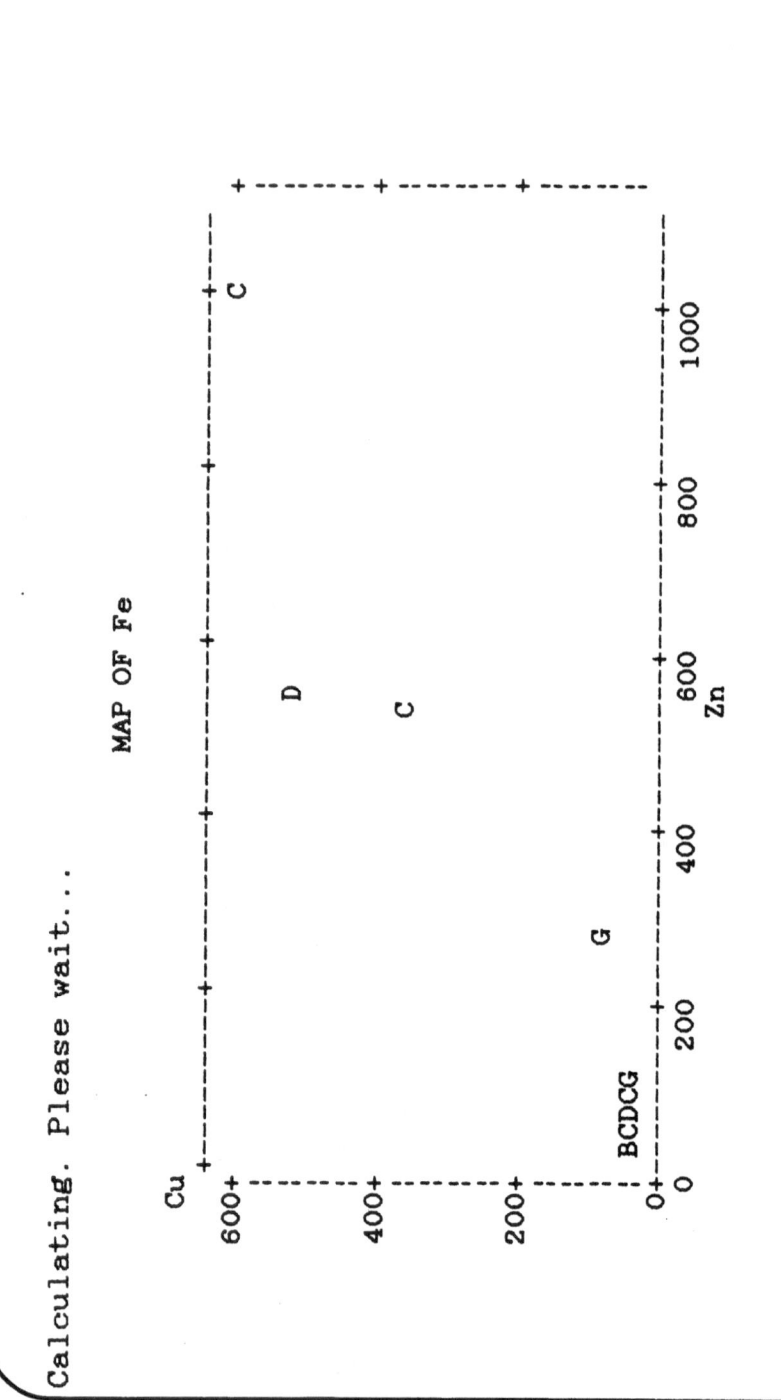

```
LEGEND
A =    1.00   B =     2.00   C =     3.00   D =     4.00   E =   5.00
F=     6.00   G=      7.00
Press any key to continue.  (The screen will clear.)

PART 8.  PLOT ANOTHER SET OF VARIABLES?

(A) Do you want to plot another set of variables? (Y/N): N
```

Error Messages

For this book's programs, Table A.1 lists error messages written to supplement the standard Microsoft BASIC messages. Although they are intended to be clear as stated, some additional explanation, keyed to the table numbers, may help you.

1. An often-made mistake is pressing the <ENTER> key without making an entry. Since every entry in this set of programs must be a definite response – for example, a variable name or a class interval – this message allows you to recover from this mistake.

2. You need to type **Y** for *Yes* or **N** for *No* (or the equivalent lowercase letters). This message prompts you to recover from accidentally typing something else.

3. A zero or negative value is inappropriate for several entries – for example, the width of a class interval for a frequency distribution.

4. A title containing too many characters would require too much space to print.

5. A negative value is inappropriate for several entries – for example, the constant to add to observations for the lognormal distribution in program 6-1. Compare message **3**.

6. See the list of ASCII characters in your computer manual or any BASIC textbook for those that cannot be printed.

7. If a number is required, it must be entered as a numeral – for example, 1, not one.

8. For some variables – for example, station name or number (Program 2-1) – you are restricted to a specific number of characters, ten in this instance.

9. You cannot specify more variables than appear in a data file.

TABLE A-1
List of Error Messages

Number	Message
1.	You evidently pressed the <ENTER> key without making an entry.
2.	You answer must be **Y** or **N**.
3.	Your value must be positive.
4.	Your title is too long.
5.	Your value must be zero or positive.
6.	You have entered the number of an unprintable ASCII character.
7.	Your entry is not a number.
8.	You have entered too many characters.
9.	You have entered too many variables.
10.	Your specified variable is not in the file.
11.	Invalid decimal places.
12.	You have entered too many title lines.
13.	You cannot numerically process a string variable.
14.	The file you specified already exists.
15.	You specified the same file name before.
16.	You have used this symbol before.
17.	You have specified an empty file.
18.	Your entry is too big.
19.	Your answer must be **A** or **D**.
20.	Your value must be positive or zero.

10. You have specified a variable that does not exist in your data file, perhaps by typing "**zn**" instead of "**Zn**", or perhaps by typing "**U**" instead of " **U**". (You must precede the symbol for a one-character variable name by a blank space.)

11. The number of decimal places is restricted to a certain range in order to have space to print them.

12. Title lines are restricted to a certain number for convenience in printing them.

13. The only variables that you work with as numbers are the x- and y-coordinates and the geochemical variables. If any of the other variables turn out to be numerical, perhaps a numerical code for rock type, you can process them as explained in Program 3-1.

14. This message prompts you to select another name. The program will not permit you to overwrite an existing data file.

15. You cannot use this particular file name, because you already have a file with this name.

16. This message keeps you from using the same symbol to identify two columns in a data file.

17. This message tells you that you have asked to process data from a file that contains no data.

18. Too large an entry cannot be handled by this program.

19. In selecting an ascending or descending frequency distribution, your responses must be **A**, **D**, or the equivalent lowercase letters.

20. Compare message **5**.

21. Your response to the menu must be a number between 1 and 12.

22. Therefore the program cannot calculate a scale.

23. Compare message 21.

ASCII Characters

Every standard character that can be displayed on your screen or printer using the BASIC language is numbered according to the American Standard Code for Information Interchange (ASCII). Unfortunately, some characters have different numbers according to whether you use the screen or printer; moreover, the numbers may vary according to the make or model of printer. For this book, the issue arises only for Programs 5-1 and 10-1, which plot data. To simplify the programming and reduce the number of questions and answers in these programs, I have adopted as the one you see on the screen a character set that will print the same character on most, if not all, printers. This set is not as attractive as some that are available.

Modifying these programs to obtain more attractive character sets for particular printers is not difficult to do, if you wish. For instance, the double lines surrounding the menu as displayed on the screen, print as other characters on an Epson Spectrum LX-80 printer. To print the somewhat different border surrounding the menu (Figure 1.2), I made these changes in BASIC program MENU.BAS: Line 1080, 186 to 124; line 1100, 201 to 124, 205 to 45, and 187 to 124; line 1260, 200 to 124, 205 to 95, and 188 to 124.

Index

177